D0153260

JAMES C. G. WALKER
The University of Michigan

Earth History

THE SEVERAL AGES OF THE EARTH

Jones and Bartlett Publishers, Inc.
Boston Portola Valley

Editorial offices: Jones and Bartlett Publishers, Inc., 30 Granada Court, Portola Valley, CA 94025.
Sales and customer service offices: Jones and Bartlett Publishers, Inc., 20 Park Plaza, Boston, MA 02116.

Library of Congress Cataloging in Publication Data

Walker, James C. G. (James Callan Gray)
Earth history: the several ages of the earth.

Bibliography: p.
Includes index.
1. Earth sciences. 2. Paleontology. 3. Human evolution. I. Title.
QE26.2.W35 1986 550 85-23958

ISBN 0-86720-022-7

ISBN 0-86720-023-5 (pbk.)

Cover illustration courtesy of NASA.

STAFF FOR THIS BOOK
Editing and design: Elizabeth W. Thomson
Illustrations: John Hamwey
Cover design: Rafael Millán

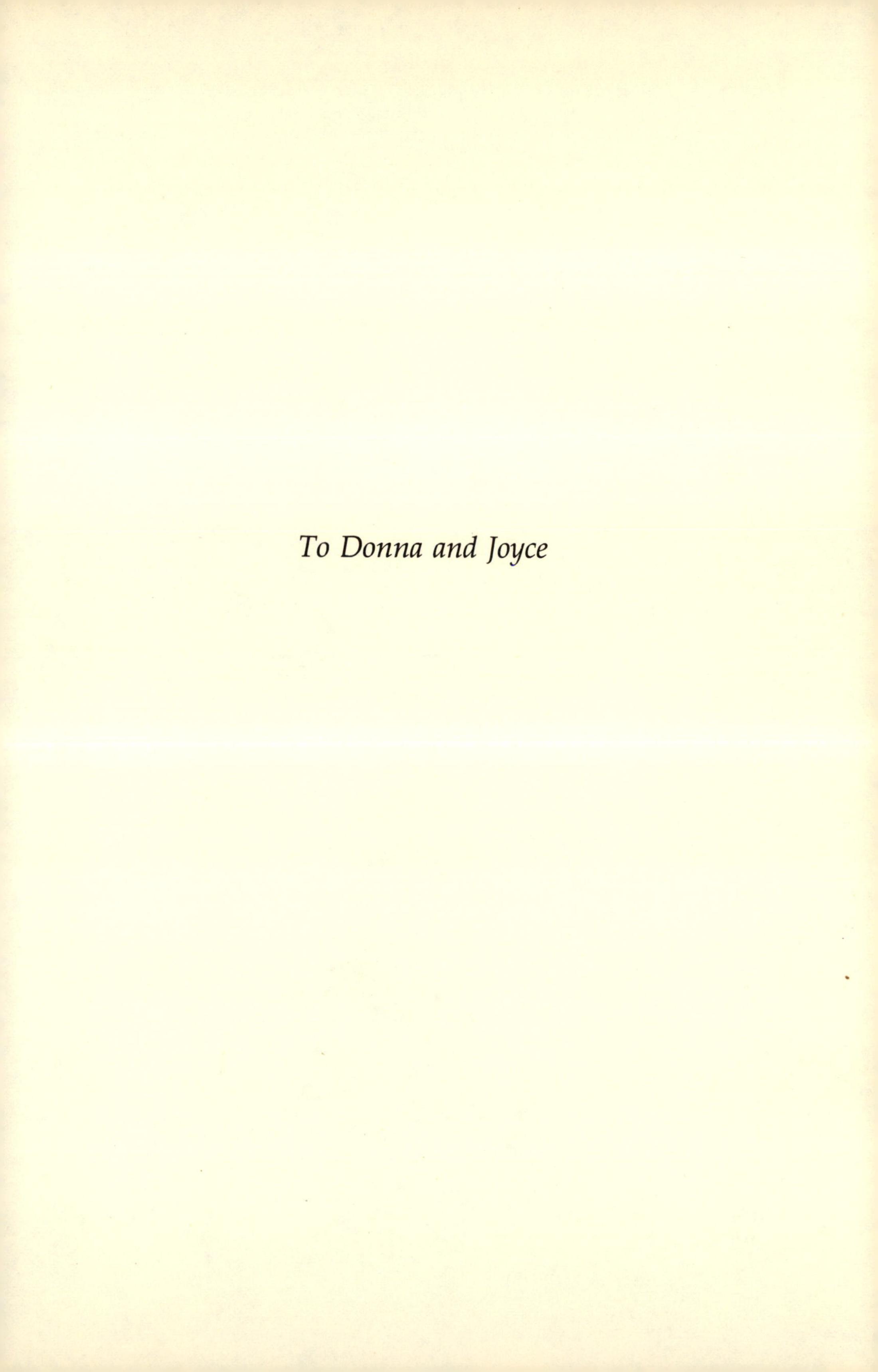

To Donna and Joyce

Preface

OURS IS a wondrous world, with mountains and rivers, forests and plains, oceans and atmosphere, and a host of living creatures. Has this diverse world always existed? If it has not, what was it like before, and how did it get to be the way it is?

No, our world has not always existed in the form that we know. Five billion years ago there was no Earth. Four and a half billion years ago there was no life on Earth. Three and a half billion years ago there were no continents on Earth. One billion years ago there were no plants or animals, only microbes. And ten million years ago there were no human beings. By what means has the complex and beautiful world of today developed from lifeless and chaotic beginnings?

This book is an attempt to answer this question from the point of view of a natural scientist. Although this account of earth history is scientific, it is a story that can be understood by readers with no scientific training. My emphasis on evolutionary processes and on the mutual interaction of life and the environment is unusual in a book on earth history, so I hope that students and scientists as well may find much here that is new to them. In my view, life is the central element of earth history. Life sets Earth distinctly apart from all other known planets. As an illustration of how life has enriched the history of our planet, consider that the physical interactions of fundamental particles have produced just under 100 chemical elements; chemical interactions have combined these elements into approximately 500 naturally occurring minerals; but biological interactions have yielded some 250,000 different species of flowering plants alone. A theme that runs through this book is the continuing interaction between life and its nonliving environment. I have sought to explain how this interaction has directed the course of evolution of both the living and the inanimate worlds.

The book is organized to respond to the key questions about earth history posed in the first chapter. Chapters 2–5 deal with the origin of land, atmosphere, oceans, and life. Then Chapters 6–11 describe the interactions of life and the environment and the effects of these interactions on earth history. Chapters 12–16 discuss the origin and evolution of mankind and its effect on Earth's continuing history. Suggestions for further reading follow each chapter.

During the writing of this book I received much help, both scientific and editorial, from Andrew Knoll of Harvard University; James Valentine of the University of California at Davis; Andrew Watson, Phillip Gingerich, and Herschel Weill of The University of Michigan; James Kasting of the Ames Research Center of the National Aeronautics and Space Administration; Lynn Margulis of Boston University; David Chapman of the University of California at Los Angeles; and my mother, Joyce Walker. To all of them I express my thanks. I am also grateful to Dianna Nickolas, Sandy Hicks, Sally Huetteman, Susan Griffin, and Matha Moon for their secretarial assistance and to Donna, my wife, for preparation of the index.

James C. G. Walker

Contents

CHAPTER 1 *Key Questions*

THE PRESENCE of life makes Earth the most complicated and interesting planet in the solar system. Indeed, the evolution of life is the only reason that we are here to wonder about earth history. In wondering what makes biological evolution central to earth history, we ask these fundamental questions: How has life affected the physical and chemical environment, and how has the environment affected the evolution of life? The answer to both questions is, a lot. The impact of life on its environment and the impact of the environment on life are recurrent themes in earth history and in this book.

Geology is unusual among the natural sciences because of its deep concern with history. This concern imparts to the science its unique mixture of observation, theory, and historical fact. Observation and theory inform us about how things work and what is possible. History selects the events that actually happened from those that are theoretically possible. Neither aspect is subordinate in geology. Because the historical record is incomplete and obscure in many places it must be read and interpreted in the light of theoretical understanding. But theoretical understanding, in earth history, acquires validity only if it can be tested against the historical record. This book is about the principles and processes of earth history, but it is not a theoretical book. Objective historical facts, the *what* and *when* and *where* of earth history, are as much a part of the story as the *how* and *why*.

The body of theory that illuminates earth history is drawn

from the disciplines of physics, chemistry, and biology, and is adapted and extended by the additional insights that the historical perspective affords. Objective historical facts—the geological archives—are preserved only in rocks. There are three different kinds of rocks, all of which contribute to the story. Igneous rocks are formed by cooling and crystallization from an initially molten state. Sedimentary rocks consist of material that was eroded from pre-existing rocks, carried away by wind or water, and then deposited either on land or under water. Metamorphic rocks are formed when either igneous or sedimentary rocks are subjected to temperatures and pressures high enough to cause chemical and structural changes.

Rocks preserve a record of the conditions under which they formed that is much more informative than the broad classification just outlined. All rocks are mixtures of minerals, crystalline solids with a definite chemical composition, or severely restricted ranges of compositions, and with definite and distinctive crystal structures. By identifying these minerals and examining their associations and interrelationships together with the textures of igneous and metamorphic rocks, scientists can deduce the provenance of the rock-forming material, as well as the temperature, pressure, and conditions of stress under which the rock was formed. The minerals of sedimentary rocks preserve information on the source of the material of which they are composed as well as on the chemical conditions under which they were deposited. The textures of sedimentary rocks can indicate whether they formed in still water, in turbulent water, or on land. Structures such as ripple marks and mud cracks provide still more information on the environments in which the rocks were deposited.

Most important, sedimentary rocks preserve fossils, the record of ancient life. Most fossils consist of hard parts such as shells, teeth, and bones. Exceptionally favorable environments—soft mud for example—have occasionally preserved impressions of the soft parts of organisms or even, as in petrified wood, fossils in which the organic matter has been replaced by mineral matter with much of the original structure preserved. Trace fossils are structures such as the tracks and burrows left by organisms in sediments that were at one time soft. Even bacteria and other microbes have been preserved as microfossils, visible only under the microscope. Useful information can also be gained from chemical fossils—organic compounds uniquely associated with life—that are occasionally preserved in sedimentary rocks.

By identifying some of the organisms that lived at various times in the past, the fossil record gives meaning to the question, How has life developed? Much of this book is devoted to the answer.

Because we are dealing with history, questions about time are also pertinent. What is the temporal framework of earth history, and how are dates established? A combination of physical and chemical evidence, much of it derived from meteorites, reveals that the Solar System, including Earth, formed about 4.6 billion years ago. The oldest rocks yet discovered on Earth, however, date from about 3.8 billion years ago, so there is no direct record of the first 800 million years of earth history (Figure 1–1). Rocks of decreasing age become increasingly abundant, but a major distinction exists, in geologic time, between the most recent 570 million years, called the Phanerozoic, and all of preceding earth history, called the Precambrian. Well-preserved fossils are abundant in sedimentary rocks deposited under water during the Phanerozoic. They provide a convenient means of distinguishing between rocks of different ages, because the nature of the fossils has changed with time as life has evolved. Studies of the succession of fossils in layered sequences of rocks have enabled geologists to develop a relative time scale in which the Phanerozoic is divided into major eras, Paleozoic, Mesozoic, and Cenozoic, and the eras are further divided into periods.

By and large, the boundaries between rocks laid down in different periods are characterized by distinct changes in the assemblage of fossils. Such a boundary marks the time when some species of organisms disappeared from the fossil record (became extinct) and new species made their first appearances. These changes in the fossil biota are particularly marked at the boundaries between the eras. The fauna of the Paleozoic, the era of ancient life, was dominated by invertebrate animals like shellfish, sponges, jellyfish, and corals. The Mesozoic was the age of reptiles, including the dinosaurs. The Cenozoic (modern life) is the era when mammals and flowering plants achieved their present abundance and diversity.

The time scale based on the fossil record of evolving life orders sedimentary rocks by their ages relative to one another, but provides no direct information about their absolute ages. The times of formation of igneous and metamorphic rocks, however, can be determined absolutely from careful measurements of the radioactive elements they contain and of the products of radioactive decay. The relative ages of igneous rocks and nearby sedimentary rocks can frequently be determined from the way they are related in space. Igneous rocks are younger than the sedimentary rocks over or through which they once flowed. They are older than any sedimentary rocks deposited on top of them. Thus a combination of geological field work and radiometric dating has provided the absolute ages of the different periods of time in the Phanerozoic.

Life originated more than 3.5 billion years ago, long before the beginning of the Phanerozoic, but it consisted just of microbes with soft bodies that left few distinctive fossils. In the absence of an abundant fossil record, clearly changing with time, Precambrian chronology depends almost entirely on radiometric dating, and not enough dates are available to establish securely the ages of all Precambrian rocks. Except in the Late Precambrian, moreover, the fossil record lacks the clearly defined changes that mark the turnover of species, so it is far from obvious how Precambrian

Figure 1–1. Geologic time.

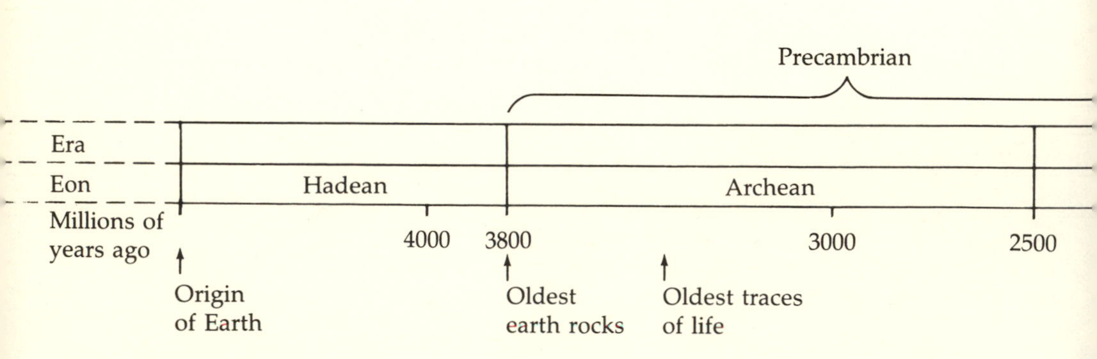

time should be subdivided. The convention that I use is shown in Figure 1–1.

Within this broad temporal framework I can pose more of the major questions of earth history in the order, roughly chronological, in which they are addressed in this book. How did Earth form, and what were the environmental conditions on the early (prebiological) earth? Because there is no geological record covering this period of earth history, the answers have to be largely theoretical. Physical and chemical reasoning is used to develop an account of what might have happened. The success of this account can be judged by how well it explains what is known of the Solar System today, including the sun, planets and their satellites, asteroids, meteorites and comets, and of course, the earth, particularly the earth as we believe it was when life originated.

How did life begin? The origin of life presents the tightest constraint we have on conditions on the early earth. It is an historical fact, albeit of uncertain date. Conditions must have been such as to permit this event to occur. Just what this means about the physical and chemical conditions at the beginning of geolog-

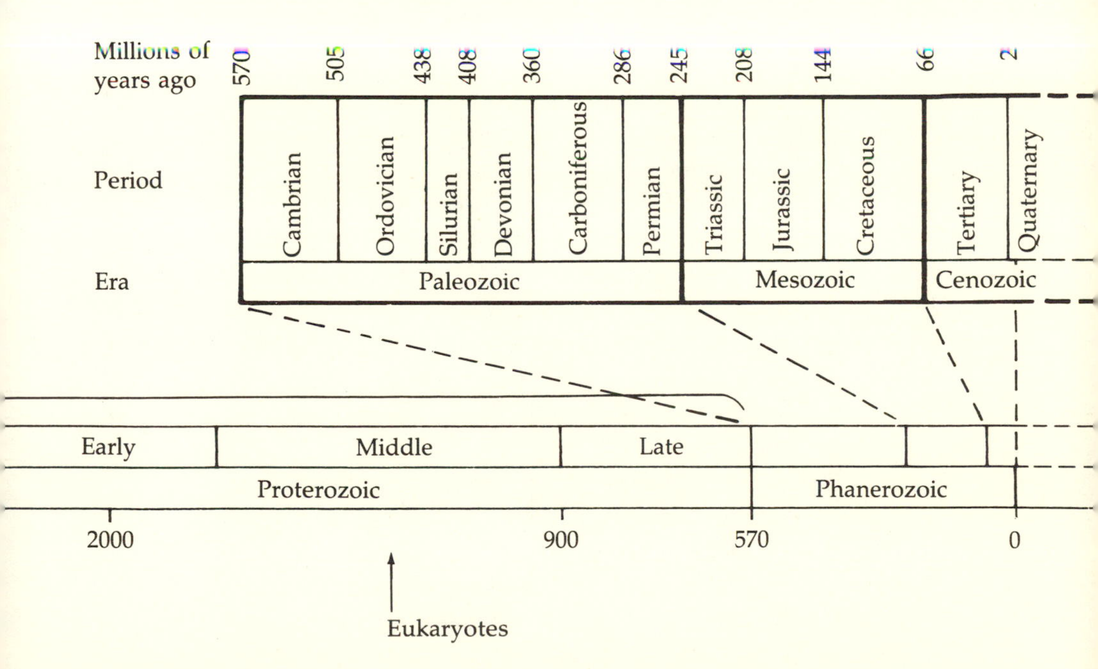

ical history remains debatable. One point, however, is clear. The prebiological earth had no free oxygen in its atmosphere. Laboratory studies of the chemical synthesis of the basic organic building blocks of life reveal that they cannot form in the presence of oxygen. Theoretical studies of the composition of primitive atmospheres show that oxygen should have been absent. And biological studies of the simplest of modern microbes that are believed to resemble the earliest forms of life show that they cannot survive in an atmosphere containing oxygen.

RADIOMETRIC DATING

Certain isotopes (atomic nuclei of a chemical element with a particular mass) undergo spontaneous radioactive decay that changes their chemical nature. The rate of this decay is a property of the isotope, and is unaffected by external factors such as temperature, pressure, or the chemical compound in which the isotope occurs. The rate is typically stated in terms of the half-life, the time required for half of the nuclei in a given sample to decay. The radiometric method of dating rock samples depends on measuring the abundance of a radioactive isotope and of the element that is produced when the isotope decays. The relative abundances of the two indicate how many half-lives have passed since the rock was formed.

For example, all naturally-occurring potassium, a chemical constituent of many common minerals, contains a small proportion of radioactive potassium-40. This isotope decays with a half-life of 1.3 billion years to form either calcium-40 or argon-40. Argon-40 is formed in 11% of the decays. Because it is a gas it is driven out of the mineral when that mineral is heated or melted. A freshly solidified igneous rock therefore contains no argon-40. With the passage of time argon-40 accumulates in the rock as a result of the decay of potassium-40. The abundances of potassium-40 and argon-40 in an igneous rock therefore tell how long ago the rock solidified (Figure 1–2).

This is the essence of the potassium–argon method of dating. Other methods that are similar in concept are uranium–lead, rubidium–strontium, and neodymium–samarium. Each method has different advantages and disadvantages as well as different potential sources of error. More than one method can be used on important rocks to provide added confidence in the results.

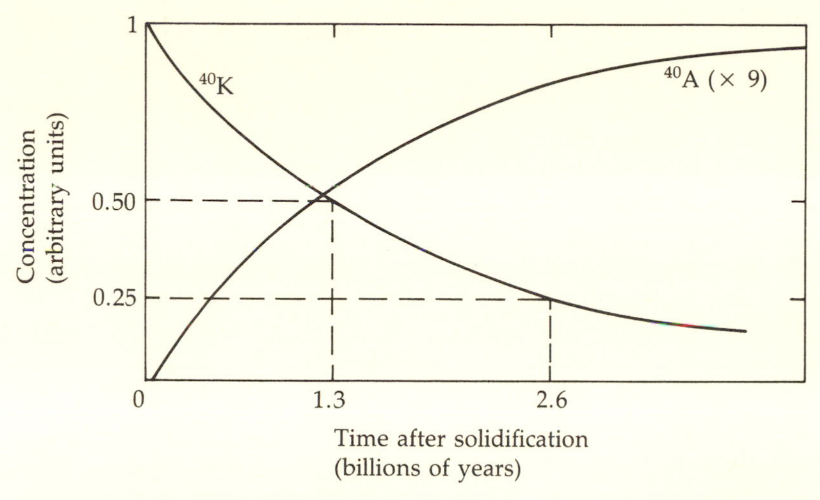

Figure 1–2. Change over time in abundance of potassium-40 and argon-40 in an igneous rock.

All the organisms we see around us every day—animals, plants, and fungi—are composed of collections of cells specialized to serve different functions within the total organism. All the cells in these multicellular organisms are known as eukaryotic cells. The eukaryotic cell has its genetic material packaged within a nucleus, and it contains specialized internal organs (called organelles) enclosed by membranes. Each kind of organelle usually has a discrete biochemical function (Figure 1–3). In contrast to the eukaryotic cell is the distinctly smaller prokaryotic cell, the cell found in bacteria. The prokaryotic cell contains no nucleus and few internal organelles. Its genetic material is not enclosed by a membrane but is dispersed through the protoplasm of the cell. Most biochemical functions are performed on internal membranes of the cell wall. The simpler structure of the prokaryotic cell suggests that it is more primitive than the eukaryotic cell, and the fossil record supports this notion. Fossils of apparently prokaryotic microbes have been found in rocks as old as 3.5 billion years. Nevertheless, the oldest eukaryotes yet discovered, tentatively identified by their size, are only 1.4 billion years old, and some paleobiologists argue that this age is too great.

The interesting point is that essentially all organisms composed of eukaryotic cells are obligate (compulsory) aerobes; that is, they can survive only in an environment rich in free oxygen. The prokaryotes, by contrast, collectively exhibit a full spectrum

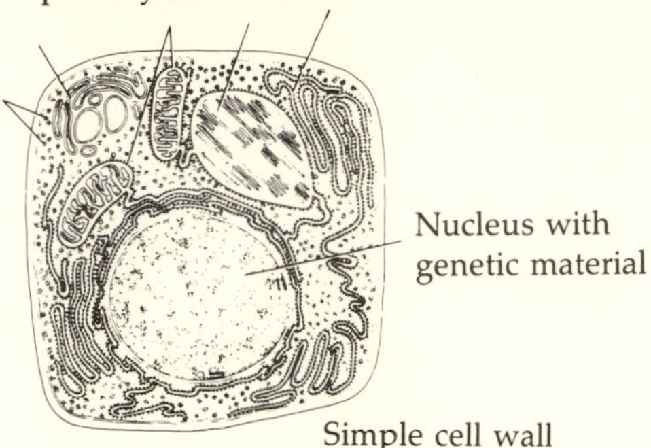

Figure 1–3. Structure of prokaryotic and eukaryotic cells. (Adapted from L. Margulis, *Early Life,* p. 3. Jones and Bartlett Publishers, Inc., Boston, 1982.)

of responses to oxygen. Some are obligate anaerobes; they are killed by exposure to oxygen. Others are obligate aerobes. Still others need or can tolerate some oxygen in their environment but not as much as there is in today's air. There are even prokaryotes that are indifferent to the amount of oxygen to which they are exposed. Some prokaryotes evidently evolved in and became adapted to an environment without oxygen, whereas the eukaryotes originated and evolved under aerobic conditions. Oxygen must have been abundant on Earth by the time of origin of the eukaryotic cell, perhaps 1.4 billion years ago. It was absent at the time of origin of life, more than 3.5 billion years ago. Some time between these dates the atmosphere became aerobic.

This transition provides the most striking example of the impact of life on its environment. Free oxygen is abundant in our atmosphere only because of the activities of plants and other organisms that convert carbon dioxide into cell material and water into oxygen. Astronomical and spacecraft measurements reveal that no other planet has an aerobic atmosphere. Life has made Earth unique in this aspect, as well as in others.

Consideration of the impact of life on its environment raises questions about the workings of the biosphere, which is defined

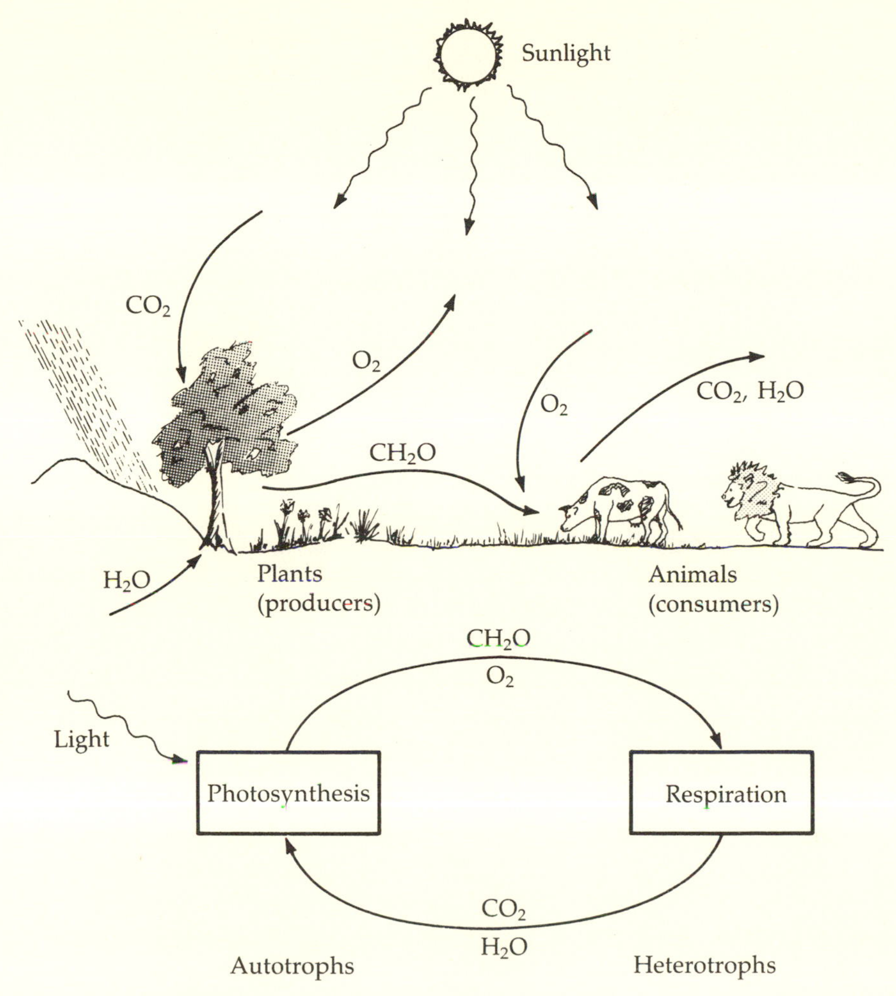

Figure 1–4. Producers and consumers.

as the combination of life and the place where it lives. What is it about the surface of the earth that makes it habitable? How do organisms and their environment cycle energy and materials back and forth in a manner that sustains this habitability without exhausting the supply of any essential elements? Part of the answers to these questions lies in a very important ecological distinction between different classes of organisms—the producers and the consumers of organic material (Figure 1–4).

The producers, of which plants are the most conspicuous examples, manufacture organic compounds—the chemical building blocks of which cells are composed—out of inorganic substances such as carbon dioxide and water. The consumers, notably animals, must be nourished by a supply of organic material that they are unable to synthesize themselves. Organisms that use carbon dioxide as their sole source of carbon are called autotrophs. Organisms that require a supply of carbon-rich organic material from which to fashion new cell material are called heterotrophs. The orderly working of the modern biosphere depends on the existence of both autotrophs and heterotrophs. On the one hand, heterotrophs rely on autotrophs for food, because only autotrophs synthesize new organic matter. On the other

FUNDAMENTAL QUESTIONS

Chapter	*Question*
1 *Key Questions*	*What is earth history?*
2 *Earth and Fire*	*How did the sun, planets, and the earth form?*
3 *Air and Water*	*Where did the atmosphere and oceans come from?*
4 *Can Air Be Anaerobic?*	*What was the atmosphere like when Earth was young?*
5 *First Life*	*How did life begin?*
6 *The Age of Microbes*	*How did life persist?*
7 *Pollution Is as Old as Life*	*Why did life dump oxygen into the atmosphere?*
8 *Pollution Can Be Helpful*	*What new ways of life did oxygen make possible?*
9 *The Not-So-Solid Earth*	*How has Earth's surface changed with time?*
10 *What Came Before Plates?*	*Have the processes of change also changed?*
11 *The Biosphere*	*What is the biosphere and how does it work?*
12 *The Age of Animals and Plants*	*How and why have life forms changed?*
13 *Good Times and Bad*	*Why are there not more living species?*
14 *Species Are Not Immortal*	*What causes species to become extinct?*
15 *The Age of Mankind*	*How did humans originate?*
16 *The Age of Ideas*	*How do humans affect earth history?*

hand, autotrophs need heterotrophs to consume organic material and convert it back to the inorganic compounds that are food for the autotrophs. In the modern world, plants convert carbon dioxide and water into organic material and oxygen. Because energy from sunlight enables plants to perform this chemical synthesis, the process is known as photosynthesis. Animals do the opposite. In the process known as respiration they obtain energy from a series of chemical reactions involving organic material and oxygen. These reactions yield carbon dioxide and water. In a world without heterotrophs, plants would soon exhaust their supply of carbon dioxide.

With some understanding of the workings of the biosphere, we can consider further fundamental questions about the history of life. How and why did life evolve? What determines the direction and pace of evolution? How did mankind originate? And, keeping in mind the mutual interactions of life and the environment, what is the impact of humanity on earth history?

The world we know best, that of the present day, furnishes only a glimpse of a fabric of interdependence within the biosphere, a fabric that has been constantly changing for more than 4 billion years. The working of the biosphere cannot be fully appreciated, therefore, without consideration of the past and what is known of the changes brought about by the passage of time. The history of the earth is the story of how we got where we are today. The better we understand the principles, processes, and interactions that have determined the course of the story so far, the better we can predict how the story will continue and how it can be influenced by different courses of action.

Suggested Reading

Cloud, P. E. *Cosmos, Earth and Man.* New Haven: Yale University Press, 1978.

Dott, R. H., and R. L. Batten. *Evolution of the Earth,* 2nd ed. New York: McGraw-Hill Book Co., 1976.

Eicher, D. L. *Geologic Time.* Englewood Cliffs, NJ: Prentice-Hall, Inc., 1968.

Laporte, L. F. *Ancient Environments.* Englewood Cliffs, NJ: Prentice-Hall, Inc., 1968.

PART ONE

Origins

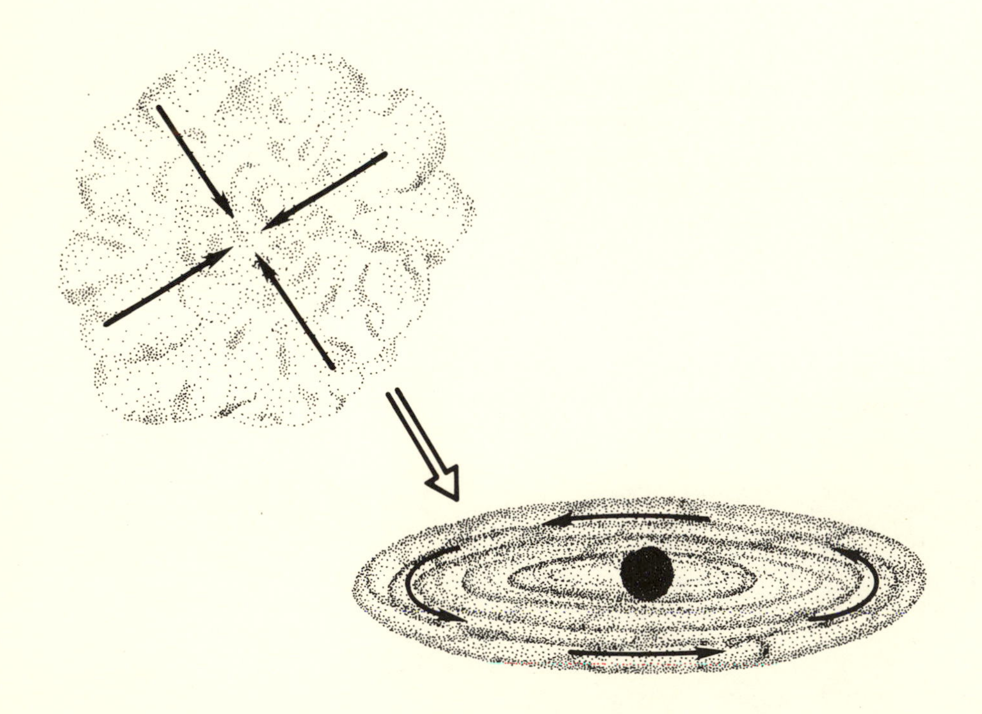

CHAPTER 2

Earth and Fire

THE SOLAR SYSTEM began about 4.6×10^9 years ago as an extended cloud of gas (mostly hydrogen), dust, and ice. Its mass was several times that of the sun and its temperature was about −263° C, just ten degrees above the absolute zero of temperature at which the random thermal motion of atoms and molecules ceases (Figure 2–1). The force of gravity caused the different parts of this cloud of gas and dust to move toward one another, so the cloud contracted. The gravitational attraction was weak at first, when different parts of the cloud were widely separated, but it grew rapidly as the cloud contracted (the gravitational attraction between objects varies inversely as the square of the distance between them). The rate of contraction therefore increased as the cloud decreased in size until contraction became collapse.

Angular momentum prevented all of this material, called the solar nebula, from collapsing into a central star destined to become the sun. The original cloud of gas and dust was rotating slowly with respect to the rest of the material in the Galaxy. Angular momentum (proportional to the rate of rotation multiplied by the square of the radius) was conserved during contraction. The rate of rotation of the cloud therefore increased as its radius decreased (Figure 2–2). In just the same way figure skaters spin faster when they bring their arms to their sides.

The increasing rate of rotation introduced another force to influence the motions of the material of the nebula. This is the centrifugal force we feel when we turn a sharp corner in a car. A

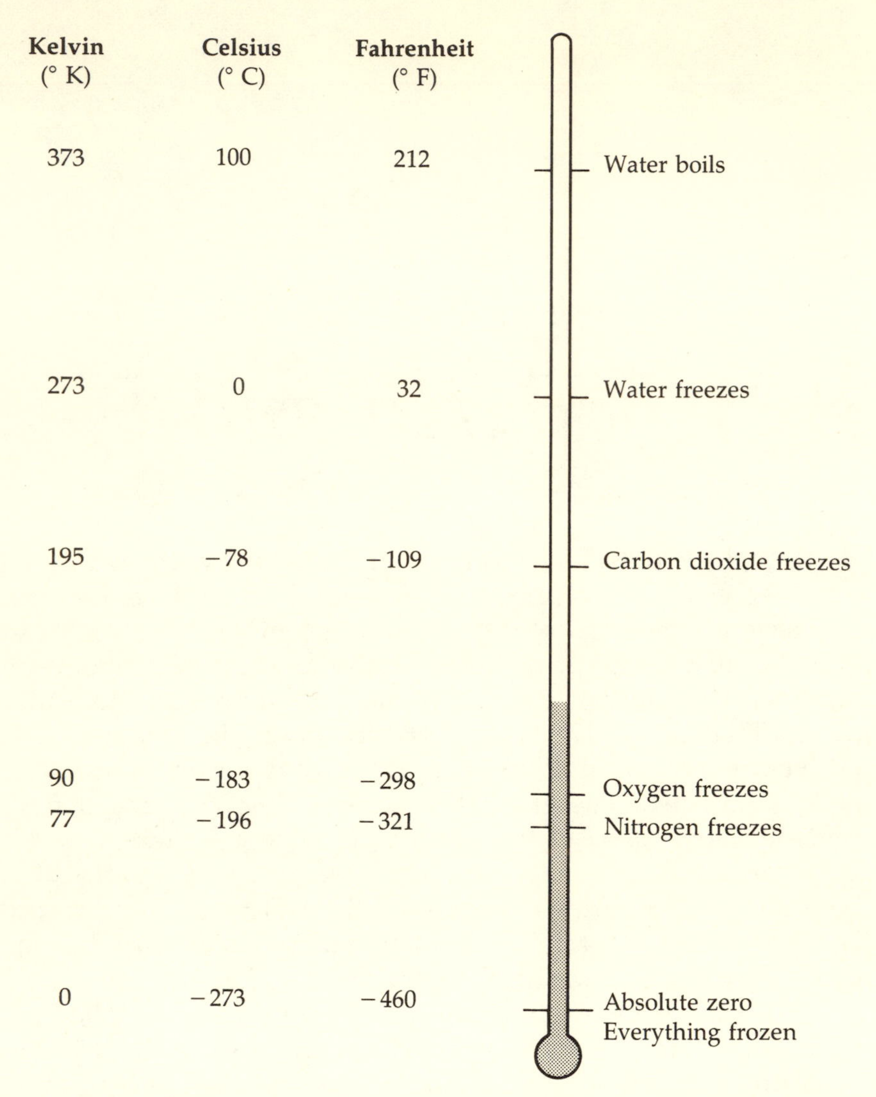

Figure 2–1. The Kelvin, Celsius, and Fahrenheit temperature scales.

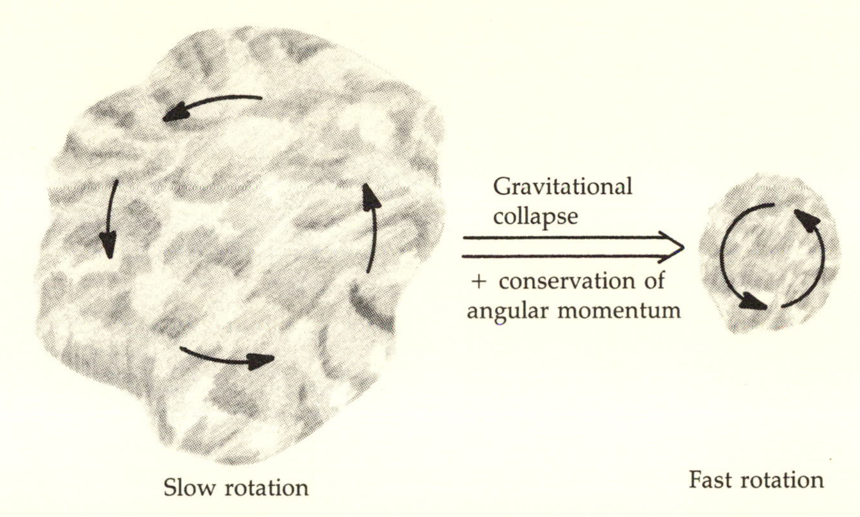

Figure 2–2. Conservation of angular momentum in the collapsing solar nebula.

planet is maintained in its orbit by a balance of the gravitational attraction between the planet and the sun with the outward centrifugal force caused by the planet's revolution around the sun (Figure 2–3). Centrifugal force increases with the rate of rotation (the angular velocity).

As the nebula contracted, then, conservation of angular momentum caused the angular velocity of some of the nebular material to become so great that centrifugal force balanced gravitational force. Material to which this happened was left behind by the contraction of the cloud to settle into nearly circular orbits

GRAVITY

From 16th century data on the motions of the planets, Sir Isaac Newton deduced that there is an attractive force between two bodies of matter that is proportional to the product of the masses of the bodies and inversely proportional to the square of the distance between them. This force—gravitational attraction—keeps the stars of the galaxy from spinning off into space and keeps the planets in orbit around the sun. It also holds us and our ocean at the surface of the earth and concentrates the gases of our atmosphere near the ground.

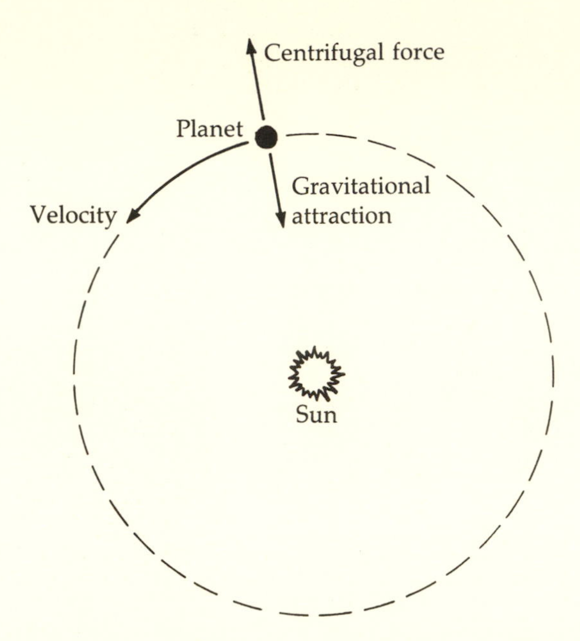

Figure 2–3. The balance between centrifugal force and gravitational attraction.

revolving around the center of the nebula. The result was that what was originally a diffuse cloud turned into a disk, rotating almost like a bicycle wheel, with a dense ball of gas at the center that was the ancestor of the sun (Figure 2–4).

As the volume of the nebula decreased, the gas it contained was compressed. Compression causes the temperature of a gas to rise. This is why a bicycle pump gets hot to the touch. The reverse effect, a fall in temperature, occurs when a gas expands. Thus, the air that escapes from a tire when the valve is released is cold. The material that suffered the most compression and therefore achieved the highest temperature was that at the center of the ancestral sun. This material reached temperatures so high that its outward pressure, which is proportional to temperature, was able to support the weight of the rest of the material in the ancestral sun (Figure 2–5). Thus, the gravitational collapse of the ball of gas was halted when the outward pressure was equal to the inward gravitational force.

Compression of the material at the center of the ancestral sun continued, however, as more and more material from the

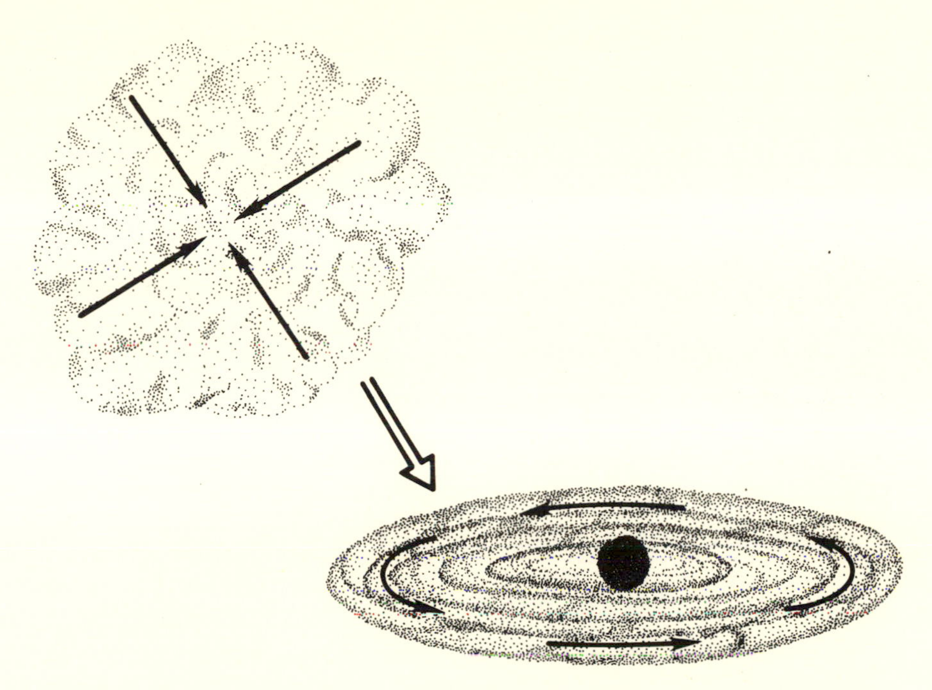

Figure 2–4. In the gravitational collapse of the solar nebula, the collapsing cloud became a rotating disk.

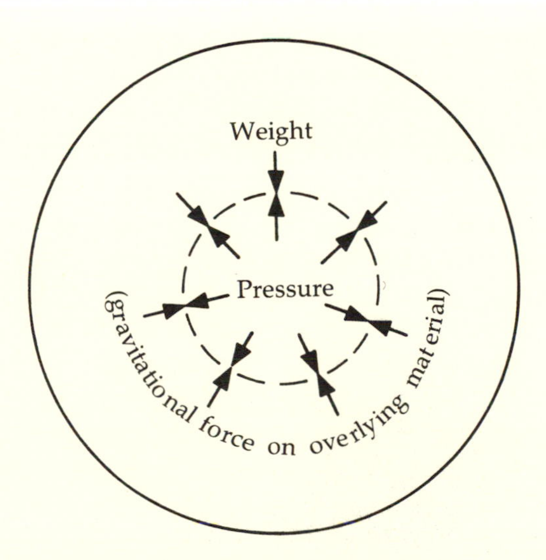

Figure 2–5. Compression of the sun, showing the balance between pressure and gravity.

nebula was sucked inwards by gravity. Temperature therefore increased while pressure also increased to support the increasing weight of the overlying layers of solar material. The molecules of a gas move faster as temperature increases; their average speed is proportional to the square root of the absolute temperature. The violence of the collisions between molecules also increases. The first effect of increasing temperature on the hydrogen gas of the ancestral sun was the disruption by collisions of hydrogen molecules (H_2) to yield a gas made up of individual hydrogen atoms (H). The hydrogen atom consists of a negatively charged electron oscillating around a positively charged proton. Higher temperatures led to collisions between atoms sufficiently violent to strip the electrons away from the protons, converting the interior of the sun to a gas composed of protons and electrons. Like charges repel one another, so very high temperatures (about 10,000,000° K) were needed before collisions became sufficiently violent to permit protons to react with one another. Eventually compression produced such temperatures in the center of the ancestral sun, and protons began to combine in a nuclear reaction that converts four protons into the nucleus of a helium atom. The mass of this nucleus is less than the mass of the four protons that reacted to produce it. According to Einstein's famous equation, $E = mc^2$, the missing mass has been converted into energy (E is energy, m is mass, and c is the velocity of light equal to 3×10^{10} cm sec^{-1}).

The research effort in controlled nuclear fusion seeks to produce and sustain, in the laboratory, temperatures sufficiently high to permit this nuclear reaction to proceed. Within the ancestral sun the required temperature was produced by compression under the weight of an enormous mass of gas. The onset of nuclear reactions in its interior marked the birth of the sun as a star. The reaction of protons, which make up most of the material of the sun, to form helium nuclei has sustained the solar fire ever since.

Let us now go back to the time before the sun became a star and describe what was happening in the nebular disk while the ancestral sun was growing in size and getting hotter. There are many competing theories of the origin of the Solar System, all of them speculative. For a start, I shall describe the one that I like best, noting later where other theories differ. Compression of the

NUCLEAR REACTIONS IN THE SUN

Energy is produced within the sun by the nuclear fusion of protons (the nuclei of the hydrogen atom) to form helium nuclei. The most important process is called the proton–proton chain (Figure 2–6). Three reactions are involved in sequence. In the first, two protons combine to yield a nucleus of deuterium (also called heavy hydrogen) plus a positron and a neutrino. The second reaction involves the combination of the deuterium nucleus with a third proton to yield a nucleus of helium-3 (the light isotope of helium) and a pulse of electromagnetic radiation called a gamma ray. Finally, two helium-3 nuclei combine to form a nucleus of helium-4 (the abundant isotope) with the release of two protons. The over-all effect of this chain of reactions is the conversion of four protons into a helium nucleus with the release of energy. An alternative chain of reactions, less important in the sun, begins with a reaction between a proton and a carbon nucleus, but ends by restoring the carbon nucleus and leaving four protons combined into one helium nucleus.

Protons are positively charged particles, which means that there is an electrostatic force of repulsion between them. Unless two protons collide with very high relative velocity this force of repulsion will prevent their getting close enough to one another to permit a nuclear reaction to take place. The relative velocities of the particles in a gas depend on the temperature. Temperatures approaching 10 million degrees Kelvin are required before protons react with one another. Such temperatures persist only in the interiors of stars.

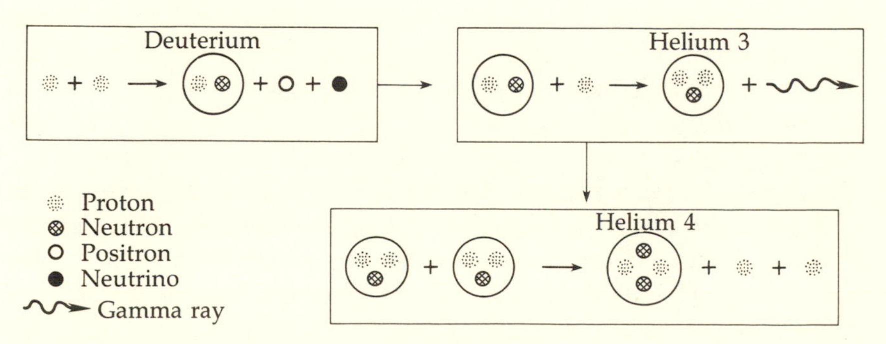

Figure 2–6. The proton–proton chain.

gas in the contracting nebula caused temperatures to rise. Near, but not in, the ancestral sun they may have reached several thousand degrees Kelvin. Farther away from the center of the disk the gas was less compressed and therefore cooler. High temperature caused the evaporation of much of the dust and ice in the nebula.

All materials evaporate (change from solid or liquid form to gas) at sufficiently high temperature. The change is a result of increasing energy, proportional to temperature, of the thermal motions of individual molecules in the material. Materials that evaporate at relatively low temperatures are said to be volatile. Examples are the gases that make up the earth's atmosphere. Carbon dioxide evaporates at 195° K, oxygen at 90° K, and nitrogen at 77° K. Thus, nitrogen is more volatile than either oxygen or carbon dioxide. Compounds that evaporate at relatively high temperatures are called refractory. Examples are the ceramic materials used to line furnaces. The high temperatures in the inner reaches of the solar nebula caused the evaporation of all but the most refractory compounds present there. Metallic iron (Fe) and the minerals corundum (aluminum oxide), perovskite (calcium titanate), and gehlenite (calcium aluminum silicate) are the compounds most likely to have survived in solid form.

MINERALS

A mineral is a naturally-occurring solid with a definite chemical composition or a very limited range of compositions. Minerals are crystalline, which means that the atoms that comprise them are bonded chemically in a systematic three-dimensional array that is characteristic of the mineral. The orderly structure of a mineral is frequently reflected in the regular shapes of individual crystals of the mineral.

The most abundant minerals on earth are the silicates, composed of silicon, oxygen, and other elements such as iron. The dominant structural component of a silicate mineral is a tetrahedron of four oxygen atoms with a silicon atom at the center. Different silicate minerals have these silicon–oxygen tetrahedra arranged in different ways with respect to one another and have different admixtures of other elements (Figure 2–7).

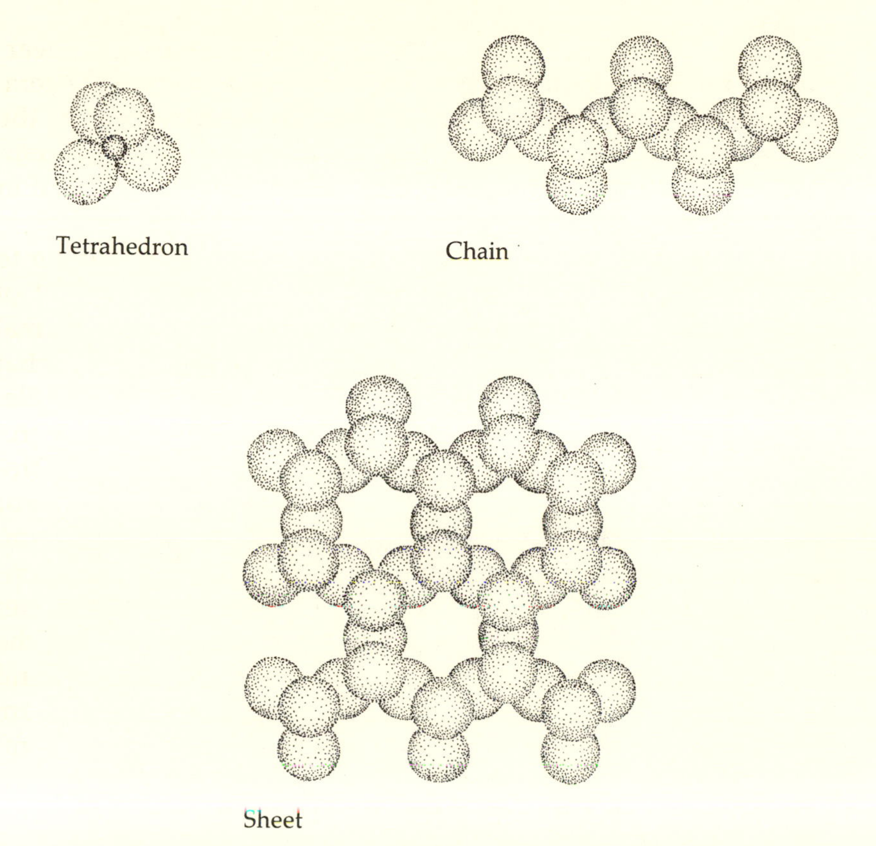

Figure 2–7. Silicon–oxygen tetrahedra in various configurations.

An additional consequence of high temperatures has already been mentioned in connection with the interior of the ancestral sun. Collisions between gas molecules at high temperature can be sufficiently violent to disrupt the chemical bonds that hold the molecules together. Thus, many of the compounds originally present in the solar nebula were broken apart into their constituent atoms or into simpler, more rugged molecular components. A combination of chemical theory and experiment makes it possible to calculate just what compounds will be present at a given temperature in a gas of specified elemental composition. Current understanding of the early history of the Solar System depends heavily on such calculations.

Hot matter emits energy in the form of electromagnetic radiation. A light bulb is an example. The primitive solar nebula lost energy steadily as a result of the emission of radiation to space. The temperature of the gas therefore began to fall once gravitational compression had been slowed by centrifugal force. As temperatures fell the atoms and simple molecules that were present at high temperatures recombined to form new chemical compounds, and the more refractory constituents began to condense. The order of condensation in a cooling gas of nebular composition can be predicted (Figure 2–8).

The most abundant early condensate was iron; common silicate minerals condensed next. When temperatures fell below about 700° K, any iron that was still exposed to the nebular gas was oxidized first to FeO and then to Fe_2O_3 (more oxygen per iron). At temperatures below about 400° K hydrocarbons began to condense, and at lower temperatures still, water entered the solid phase in the form of hydrated silicate minerals (claylike minerals containing chemically bound water).

The hydrocarbons, compounds of hydrogen and carbon such as oil or asphalt, are particularly interesting because when

CHEMICAL FORMULAS

Each chemical element is represented by a symbol of one or two letters, usually the initials of the name of the element in English or Latin. A chemical compound is represented by a combination of these symbols that indicates the proportions in which the elements are present. The water molecule, for example, combines two atoms of hydrogen with one of oxygen; its symbol is H_2O. *Carbon dioxide combines two atoms of oxygen with one carbon atom; it is written* CO_2.

A chemical reaction can be described by indicating the molecules that react and the molecules that are produced. For example, the burning of methane (CH_4), *a major constituent of natural gas, involves a reaction with oxygen* (O_2, *because the oxygen molecule contains two atoms) to produce carbon dioxide and water:*

$$CH_4 + 2O_2 \rightarrow CO_2 + 2H_2O$$

In words: one molecule of methane reacts with two molecules of oxygen to produce one molecule of carbon dioxide and two molecules of water. For each element, there are as many atoms among the products as there are among the reactants; matter is conserved.

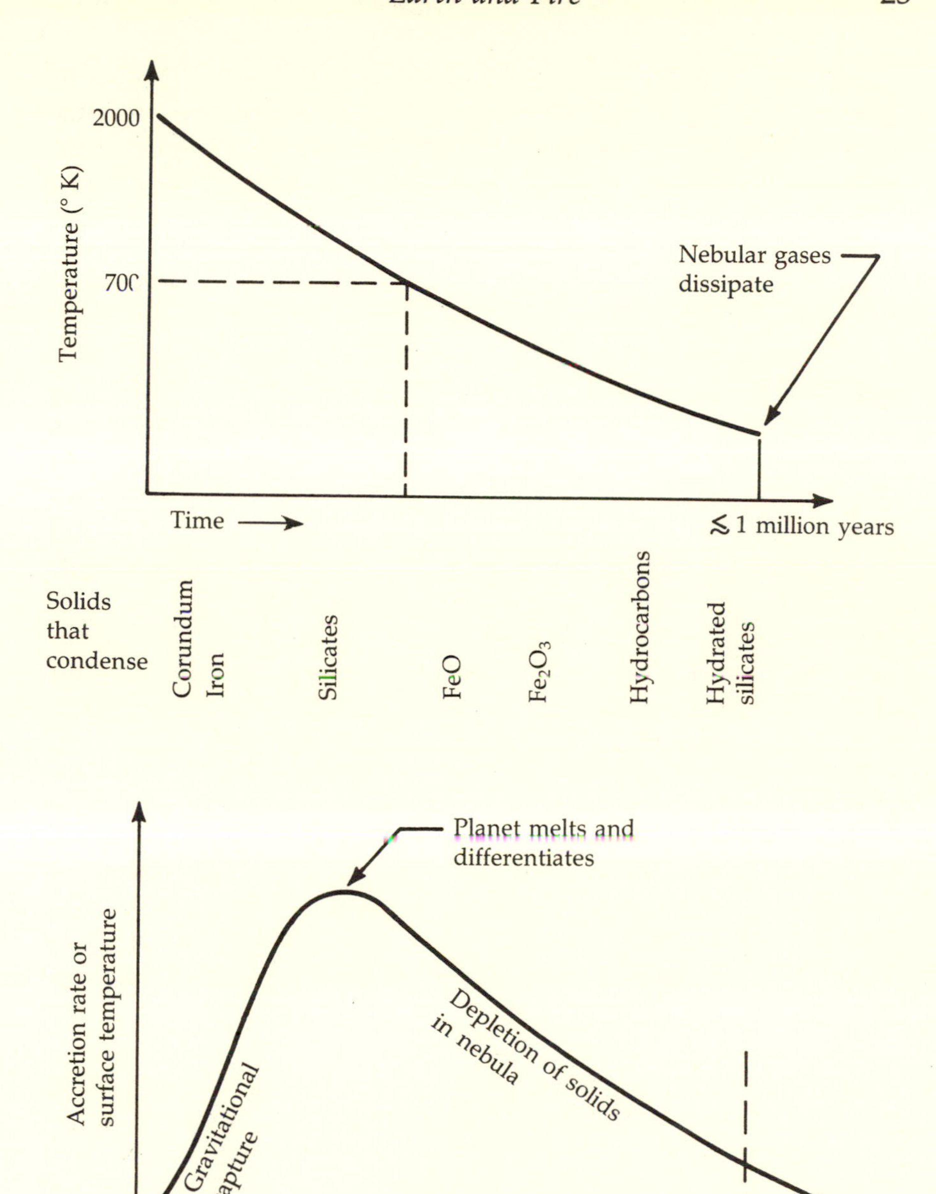

Figure 2–8. Cooling and condensation in the nebula and accretion of a planet. (Adapted from J.C.G. Walker, *Monthly Notes of the Astronomical Society of Southern Africa, 35,* 2, 1976.)

they condensed they incorporated nearly all of the carbon that entered the solid phase. The planets formed by the accumulation of solid debris, leaving gases in the solar nebula, so carbon entered the planets in the form of hydrocarbon compounds. Today this carbon is the basis of terrestrial life, and as carbon dioxide it constitutes most of the atmospheres of Venus and Mars.

The planets were growing throughout this period of cooling and sequential condensation. Solid particles near each other exerted a gravitational attraction that caused them to move toward one another through the nebular gas and to collide. Ever larger aggregations of solid material grew as a result. Because the gravitational force depends on mass, large aggregations grew more rapidly than small ones. The largest ones gobbled up their smaller neighbors to become the ancestors of the planets.

The protoplanets grew slowly when they were small because their gravitational fields were weak. Their rates of growth accelerated as they grew more massive until eventually they exhausted the supply of smaller neighbors and growth was once again slow. A curve of the rate of growth of a planet as a function of time therefore has approximately the shape of a bell (Figure 2–8). Gravitational energy was converted into heat when a solid particle fell onto the surface of a growing planet, just as heat is released when meteorites or artificial satellites fall to Earth today. The growing planets were therefore heated by the impact of the material that contributed to their growth, and the rate of heating was proportional to the rate of growth. Around the time of maximum growth, the rate of heating was probably large enough to melt the protoplanets or at least to soften their materials to a nearly fluid state. This fluidity permitted the denser constituents of a planet to settle to the center under the action of gravity, leaving light constituents to float on top. The subsequent decrease in the rate of growth permitted temperatures to fall, causing the planet to solidify while preserving a stratified internal structure consisting of a dense core of iron and nickel overlain by a less dense mantle of silicate minerals (Figure 2–9). According to this theory, therefore, Earth acquired its internal structure (shown in Figure 2–10) during the process of accretion, while the planet was still immersed in the primitive solar nebula. Any volatile materials incorporated in the protoplanet would have been driven off into the nebula by high temperatures at the time of melting.

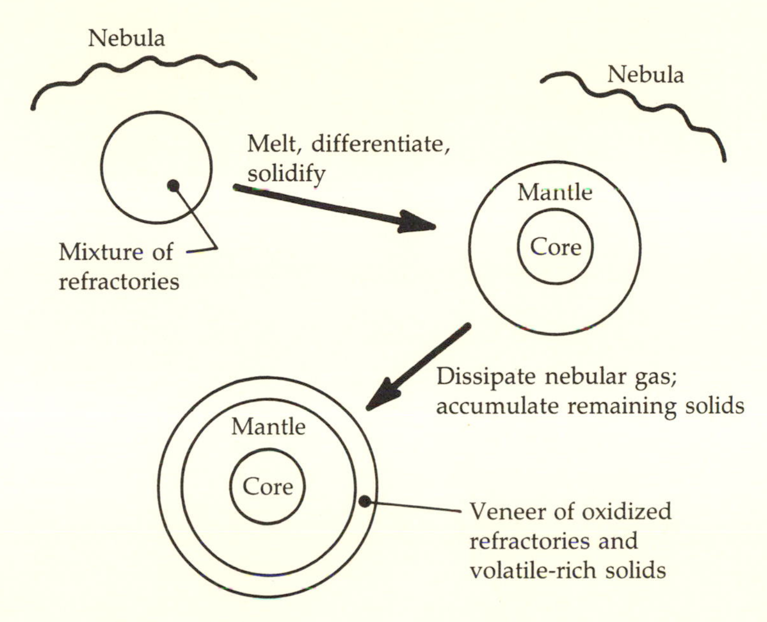

Figure 2–9. Stages of planetary growth. (Adapted from J.C.G. Walker, *Monthly Notes of the Astronomical Society of Southern Africa, 35,* 2, 1976.)

Temperatures in the solar nebula continued to fall after the planets had melted and solidified again, so new, less refractory (more volatile) solids continued to condense from the nebular gas and become available for accretion. The newly condensed material was added to the upper layers of the planets along with the remnants of refractory solids that had condensed earlier but had escaped previous capture. Condensation of the new solids ended when the gases of the nebula were blown away into interstellar space by the pressure of particle and electromagnetic radiation from the young sun. Accretion continued, however, at a rate that tapered off as the supply of solid particles was exhausted. The surface layers of the earth, including upper mantle, crust, ocean, and atmosphere, are derived from this mixture of refractory and volatile solids that accreted after melting and differentiation of the planet. Atmosphere and ocean originated during the final stages of accretion, after dispersal of the solar nebula.

For the sake of clarity, I have presented this account of the origin of the Solar System without the numerous disclaimers and caveats that it undoubtedly deserves. At present it is little more than a speculation, based largely on the ideas of K. K. Turekian

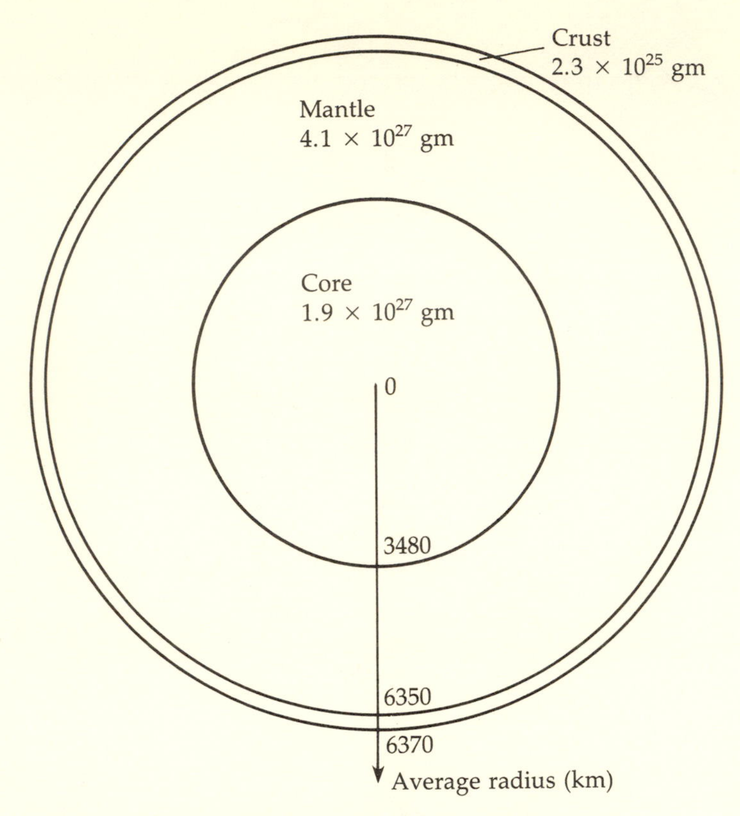

Figure 2–10. Internal structure of the earth.

and S. P. Clark of Yale University. The availability of new data concerning the moon, planets, and meteorites has stimulated vigorous research in this area. Equally plausible alternative theories are hotly defended, and all existing theories are subject to critical attack.

It seems quite clear that the inner planets formed by the accretion of a mixture of refractory and volatile solids, with the proportions in this mixture varying from one planet to another. The major disagreements with this theory are about the extent to which the temperature changes that produced these different solids occurred in time rather than in space. The theory presented above places major emphasis on a supposed cooling of the solar nebula, with refractories condensing before volatiles. An alternative theory, developed by J. S. Lewis of the Massachusetts Institute of Technology, emphasizes a decrease of temperature with distance from the sun rather than with time. According to this theory, material that condensed near the sun was refractory,

PLANETS, ASTEROIDS, AND COMETS

The objects in the Solar System can be divided into three classes according to their densities and compositions. There are rocky objects like the inner planets (Mercury, Venus, Earth, and Mars) and their satellites, as well as the asteroids and meteorites. There are objects that consist mainly of gas (principally hydrogen), such as the outer planets (Jupiter, Saturn, Uranus, and Neptune) and the sun itself. And there are objects that are a mixture of rocky material with frozen water, methane, and ammonia. Examples of these objects are the nuclei of comets and the satellites of the outer planets. Any theory of the origin of the Solar System must explain the existence of these different kinds of objects. How do the theories I have presented measure up?

The gravitational fields of the protoplanets concentrated nebular gas around them to form atmospheres with the composition of the nebula (mostly hydrogen). The greater the mass of the primitive planet the stronger its gravitational field and therefore the greater the mass of this primordial atmosphere. Probably during dispersal of the gaseous nebula these primary atmospheres were driven away from the inner planets but not from the more distant and more massive outer planets. The dispersal of the nebula therefore left the outer planets essentially as they are today, balls of gas of more or less nebular composition. The inner planets, in contrast, were left as naked balls of rock, ready to be altered profoundly by subsequent events.

Temperatures were greater in the inner portions of the solar nebula than in the outer portions during the time when solids were condensing and accumulating to form planets. Objects in the outer reaches of the Solar System therefore contain a greater proportion of volatile material, which condenses at low temperatures, than objects in the inner Solar System. This is why there are icy objects in the outer Solar System and rocky objects in the inner. A quantitative theoretical formulation of this simple idea by J. S. Lewis of the Massachusetts Institute of Technology has been remarkably successful in describing the bulk properties of the inner planets and the satellites of the outer planets.

whereas volatile-rich material formed at greater distances. Like Lewis, A. E. Ringwood of the Australian National University assumes that the change of nebular temperature with time was of minor importance. Ringwood suggests that temperature varied significantly in the direction perpendicular to the plane of the nebular disk. Volatile-rich material condensed at the cool upper and lower surfaces of the disk and then settled under the action of gravity to the hot central plane, where it accreted into planets along with locally condensing refractory material.

It is hard to imagine that Lewis and Ringwood are wrong about the decrease of temperature both with distance from the sun and with distance from the central plane of the nebular disk. The importance of these temperature changes as compared with the temporal change emphasized by Turekian and Clark depends on the relative time scales for nebular cooling, planetary accretion, and the dissipation of the nebular gases into interstellar space. If accretion was rapid compared with cooling and if substantial cooling occurred prior to dissipation of the nebula, then the planets would have accreted refractories before volatiles, as I have described. If there was little change in temperature before dissipation, then the mixture of volatiles and refractories that accreted to form a planet must have come from different parts of the nebula, probably with little change in the mixture as the planet grew. Further theoretical work on the thermal structure and cooling rate of the nebula, its dissipation, and the processes of planetary accretion will be needed to clarify these issues.

Particularly important is the mechanism and timing of the dispersal of the nebular gas. When and how this occurred is not at all well understood. It has important implications for the origin of the atmosphere, because the gravitational force of a planet immersed in a gaseous nebula would attract a substantial atmosphere composed principally of hydrogen, the dominant constituent of the nebula. A theory that has the planets accreting before dispersal of the nebula must deal with the subsequent fate of this primary atmosphere.

The present atmospheres of the inner planets appear to be secondary, in the sense that they have been formed by the release of gases from the solid bodies of the planet. This conclusion is based on a comparison of the compositions of these atmospheres with the sun. (The evidence is discussed in the next chapter.)

Some years ago there was no conflict between the ideas of a primary atmosphere of nebular composition and a secondary atmosphere of quite different composition. It was assumed that the primary atmospheres escaped into space, possibly at the time of dissipation of the nebula, long before the secondary atmospheres were driven out of the interiors of the planets, a process known as degassing. It seemed likely that degassing was delayed until an initially cold and rigid planet had been heated to a nearly fluid state by the decay of radioactive minerals. Only then, it was believed, could dense matter settle to the center and light matter rise to the top to differentiate the interior of the earth into core, mantle, and crust.

More recent studies, incorporated into the most widely discussed current theories, indicate that heating by accretion yielded planetary interiors that were hot and sufficiently mobile that internal differentiation and degassing occurred at essentially the same time as planetary formation. Earth was born in fire, not ice.

Suggested Reading

Chapman, C. R. *The Inner Planets.* New York: Charles Scribner's Sons, 1977.

Jastrow, R., and A. G. W. Cameron. *Origin of the Solar System.* New York: Academic Press, 1963.

Ringwood, A. E. *Origin of the Earth and Moon.* New York: Springer-Verlag, 1979.

Scientific American. *The Solar System.* San Francisco: W. H. Freeman and Co., 1975.

CHAPTER 3

Air and Water

THE LAST CHAPTER described the formation of the earth and the origin of its internal structure. This chapter describes the origin of the earth's atmosphere and ocean as well as its closely related veneer of sedimentary rocks.

That there is a distinction between liquid ocean and gaseous atmosphere is a consequence simply of the temperature of the surface. If Earth's surface temperature were as high as that of Venus (700° K), for example, the oceans would evaporate and the atmosphere would consist principally of water vapor. If temperatures were as low as on Mars, the oceans would freeze and water ice would be just another mineral constituent of sedimentary rocks. The distinguishing characteristic of both ocean and atmosphere is that they are composed of relatively volatile compounds. Moreover, they share a common history.

There are two contrasting views of the origin of Earth's surface volatiles, the constituents of ocean and atmosphere. One is that they are primary, in the sense that they are a remnant of the primitive solar nebula that has been retained by the earth's gravitational field from the time of formation of the Solar System. For example, the atmosphere of Jupiter, composed mainly of hydrogen, is primary. The other possibility is that atmosphere and ocean are secondary, in the sense that they were originally incorporated into the body of the earth as solid compounds from which the volatile constituents have since been released. There is little doubt that the second view is correct.

In the last chapter I suggested that all gas was stripped away from the inner planets, probably by radiations from the young sun, leaving them as airless balls of rock. Although the mechanism is speculative, the evidence for loss of a primary atmosphere is straightforward. It was presented and interpreted in fairly convincing form by F. R. Moulton as early as 1905. As shown in Figure 3–1, if the average chemical composition of the earth is compared with that of the sun, that of the earth is extremely deficient in the inert gases (helium, neon, argon, krypton, and xenon), which do not form compounds or condense at temperatures that existed within the solar nebula. The message is clear. Earth was formed by the accumulation of solids. The gaseous component of the solar nebula did not contribute significantly to the mass of the planet, and any atmosphere of nebular (solar) composition that Earth may once have possessed has been lost. Most of the material that now constitutes Earth's atmosphere and ocean must have been accreted originally in the form of solid compounds, which would have excluded the inert gases. *Viking* and *Pioneer Venus* spacecraft measurements have shown that the same argument applies to Mars and Venus.

After the dispersal of the solar nebula, the airless planets continued to accumulate the solid debris left behind in their vicinities. This debris was a mixture of refractory material that had condensed when the nebula was hot (or in hot regions of the nebula) and material rich in volatiles that had condensed late in the history of the cooling solar nebula (or in cool regions if change of temperature with time was unimportant). There are several lines of evidence that contribute to our understanding of the nature of this material. The material formed as a result of chemical reactions and condensation, which took place in the solar nebula at different temperatures. Chemical theory can be used to calculate what compounds should have formed as a function of temperature. (The results of such calculations were summarized in Chapter 2.) Iron condensed at the highest temperatures, followed by silicate minerals, iron oxides, hydrocarbons, and hydrated silicate minerals.

The compounds of most significance for the origin of the ocean and atmosphere were the hydrocarbons, compounds of hydrogen and carbon in varying proportions, as well as the water of hydration that combined with certain silicate minerals. The

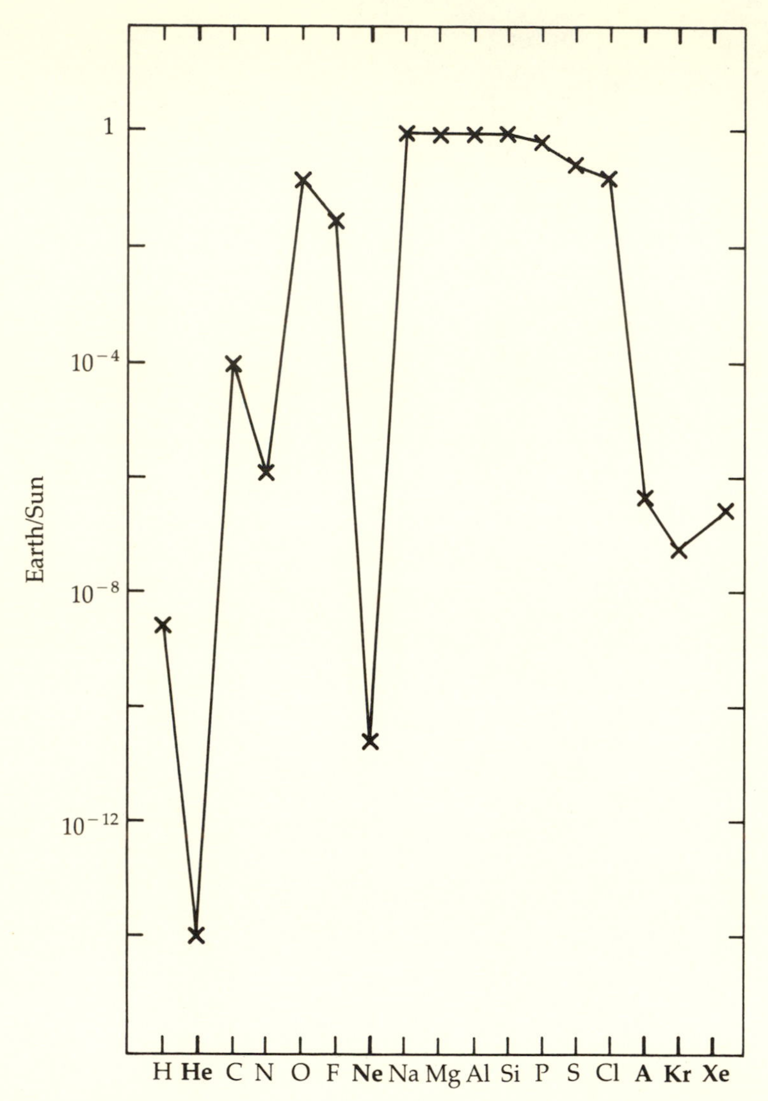

Figure 3–1. Relative abundances of different elements on Earth as compared with the sun. (The inert gases are indicated by boldface type.)

theoretical findings are supported by studies of the chemical and mineralogical composition of meteorites, which are presumably representative, to some extent, of the solid bodies that condensed in the solar nebula and accreted to form the earth. The most abundant volatile constituents of meteorites are hydrocarbons and water of hydration, as predicted by theory. Some meteorites could be surviving examples of the volatile-rich material accreted by the inner planets during the final stages of their growth, the material from which their atmospheres originated.

In addition to chemical theory and meteorites, we can use the surface layers of the earth today to gain more knowledge of what was accreted. The upper mantle and crust are composed predominantly of silicate minerals, compounds of silicon and oxygen mainly with magnesium, iron, aluminum, calcium, sodium, and potassium. Similar silicate minerals are abundant constituents of meteorites, and they are also expected, on the basis of chemical theory, to have formed in abundance at intermediate temperatures in the solar nebula. The source of the dominant chemical elements in the crust and upper mantle presents little mystery. The source of the chemicals that are now in gaseous or liquid form in the atmosphere and ocean requires more thought. In what solid form were they accreted by the earth?

The most abundant volatile compound in Earth's surface layers is water, and most of Earth's water is in the ocean. Most of this water was probably accreted in the form of water of hydration in combination with silicate minerals. Heating of the hydrated minerals either during or after accretion would have driven the water out of the interior of the earth. Additional water would have been produced by the reaction, again at high temperature, of hydrocarbons with oxygen combined in silicate minerals or with the iron oxides that earth accreted along with silicate minerals. The reaction of oxygen with hydrocarbons yields carbon dioxide as well as water.

Indeed, after water, the next most abundant volatile constituent of Earth's surface layers is carbon dioxide. This gas dominates the atmospheres of Mars and Venus, but geological processes have caused nearly all of Earth's carbon dioxide to leave the atmosphere and enter the solid earth as a constituent of carbonate minerals in the crust. The mass of carbon dioxide in the crust is about one third as great as the mass of water in the oceans. Carbonate minerals are exceedingly rare in meteorites and are not expected to have formed in the solar nebula. Evidently Earth did not accrete carbon dioxide in the form of carbonate minerals. Instead, carbon was probably accreted in combination with hydrogen as hydrocarbons. This carbon would have been released to the atmosphere as carbon dioxide when the hydrocarbons were oxidized, at high temperature, in the reaction mentioned as a source of water. An alternative suggestion, by J. S. Lewis, is that carbon in its elemental form was mixed with the metallic iron that now forms the core of the earth.

Nitrogen, the dominant constituent of the terrestrial atmosphere, is less abundant than carbon dioxide, when allowance is made for crustal carbonate, by a factor of about 100. Nearly all of Earth's nitrogen is in the atmosphere. It is not expected to have formed solid compounds in the solar nebula and is virtually undetectable in meteorites. Evidently nitrogen was accreted as a minor contaminant of the solid matter that made up the surface layers of the earth. It is likely that the presence of nitrogen on Earth results from imperfections in the chemical and condensation processes that occurred in the solar nebula. Ammonia, a compound of nitrogen and hydrogen, occasionally takes the place of potassium in the crystal lattices of silicate minerals, because a molecule of ammonia and an atom of potassium are almost the same size. A similar chemical error probably made possible the accretion of the chlorine that now comprises one of the major dissolved components of the sea (sea salt is sodium chloride, NaCl). Chlorine may have substituted occasionally for oxygen in the crystal lattices of silicate minerals. These vagrant elements, trespassing in chemical structures where they do not belong, would have been released when the rock into which they were incorporated was melted and then recrystallized.

The process by which gases were released from the solid earth to form the atmosphere and ocean is known as degassing. Two conditions had to be met in order for degassing to occur. First, the material rich in volatiles had to be subjected to relatively high temperatures in order to drive gases out of solids by melting and by chemical reaction. Second, the gases produced by high temperature reactions within the body of the earth had somehow to be brought to the surface—the outer layers of the earth had to be stirred.

When and how degassing occurred is still subject to debate. The traditional view, first developed by William W. Rubey of the United States Geological Survey in 1951, holds that degassing has been gradual; that it has taken place throughout geologic history by way of volcanoes and hot springs such as are found on Earth today. According to this view, the masses of the atmosphere and the ocean have grown steadily with time and are probably still growing. The alternative point of view, stated most forcefully by Fraser Fanale of the Jet Propulsion Laboratory in 1971, holds that there was a single short episode of degassing early in Earth's history.

There are several lines of evidence that support the notion of rapid early degassing, and this is the hypothesis that I shall adopt for further discussion. It is probable, however, that the truth lies somewhere between these limits. Most of Earth's atmosphere and ocean may have been released in an early episode of rapid degassing, but the degassing process may never have ceased altogether.

One of the most attractive features of the hypothesis of rapid early degassing is that it makes use of an abundant source of energy for the necessary heating and stirring of the outer layers. This energy was provided by the bombardment of the earth by solid material during the final stages of accretion. Degassing probably took place during the last stages of growth of the planet.

Extensive areas of Moon, Mars, and Mercury are covered by heavily cratered terrain that was formed during accretion of their surface layers. Even on Earth, where little ancient crust has survived the ravages of erosion and tectonic activity (mountain building), remnants of ancient impact craters may be found. *Apollo* studies of the moon have revealed that most lunar craters were formed more than 3.8 billion years ago. The accretion of the surface layers of the planets also appears to have been largely complete by that time.

It is probably not a coincidence that the oldest known rocks on the surface of the earth, near Isua in western Greenland, date from about this time. These rocks have been heavily metamorphosed by high temperature and pressure at some time since their formation, but at least some of them were originally laid down as sediments on the floor of the sea. This is a very significant datum. It confirms that Earth had acquired some ocean and atmosphere before it was a billion years old, although it leaves open the question of how large this ocean and atmosphere were. For purposes of further discussion I shall assume that the atmosphere grew to approximately its present mass very early in earth history as a result of degassing during the final stages of accretion.

Suggested Reading

Brancazio, P. J., and A. G. W. Cameron (Eds.). *The Origin and Evolution of Atmospheres and Oceans.* New York: John Wiley and Sons, 1964.

Lewis, J. S., and R. G. Prinn. *Planets and Their Atmospheres.* New York: Academic Press, 1984.

Taylor, S. R. *Lunar Science: A Post-Apollo View.* New York: Pergamon Press, 1975.

CHAPTER 4

Can Air Be Anaerobic?

NOT ALL of the volatiles released during degassing remained in the atmosphere or even in the ocean. Some returned to rocks as a result of weathering reactions. Weathering occurs when rocks are exposed to air and water at the surface of the earth. The exposed rocks undergo chemical changes and are broken down into fragments that may be carried away from their place of origin by wind or running water before being deposited as sediments either on land or under water. In time the sediments turn into sedimentary rocks. The weathering reactions that produce sedimentary rocks typically consume volatiles.

In fact, sedimentary rocks characteristically contain a higher proportion of volatile elements than the igneous rocks from which they were formed. The production of sedimentary rocks by the weathering of igneous rocks therefore requires a supply of volatile constituents from the atmosphere and ocean (Figure 4–1). The development of Earth's blanket of sedimentary rock was therefore a direct, though not necessarily an immediate, consequence of degassing.

Weathering reactions greatly complicate the problem of estimating the composition of the early atmosphere and ocean. It is not enough to consider just the material released by degassing; its subsequent fate must be considered as well. Time scales are also important. If degassing was rapid compared with the rate of weathering processes, a massive atmosphere may have formed initially and then decayed slowly as weathering reactions gradu-

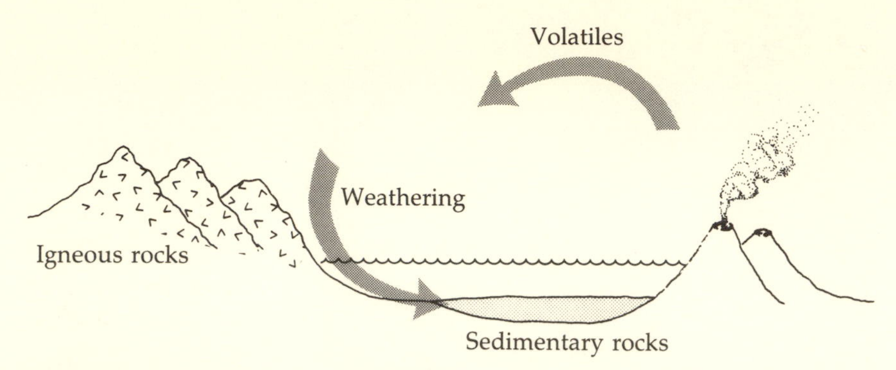

Figure 4–1. Formation of sedimentary rocks. Sedimentary rocks contain a higher proportion of volatile constituents than igneous rocks, so a source of volatiles is required for igneous rocks to weather to form sedimentary rocks.

ally restored volatiles to the solid phase. In time, a balance would have been achieved between the rate of removal of volatiles from atmosphere and ocean by weathering reactions and the rate of release of volatiles from sedimentary rocks to the atmosphere caused by heating and metamorphism (or volcanism) associated with motion within the earth. Alternatively, if weathering was sufficiently rapid, this balance between the sources of volatiles and the sinks may always have existed.

In the present state of knowledge it is hardly possible to say anything useful about the initial composition of a hypothetical atmosphere released by instantaneous degassing. Nevertheless, speculative but plausible estimates can be made for the composi-

GEOCHEMICAL BUDGETS

A budget describes the flow of material through a system. In the case of a constituent of the atmosphere, the budget describes the rate at which various processes, called sources, supply that constituent to the atmosphere as well as the rate at which other processes, called sinks, remove that constituent from the atmosphere. The amount of the constituent in the atmosphere is called the reservoir. Thus, if the sources and sinks are not in balance the reservoir must be changing with time.

tion once a balance had been achieved. The procedure is to consider the sources and sinks of each gas separately and estimate the conditions that could have brought them into balance. As an illustration I shall take carbon dioxide.

Carbon dioxide is removed from the atmosphere by weathering reactions that convert silicate minerals to carbonate minerals, compounds principally of calcium and magnesium with carbon and oxygen. The rates of the weathering reactions, and therefore the sinks for carbon dioxide, presumably increase with the amount of carbon dioxide in the atmosphere, as shown in Figure 4–2. The source, release of carbon dioxide by metamorphism and volcanoes, is independent of the amount of the gas in the atmosphere.

Suppose that the source of carbon dioxide were to exceed the sink. Gas would accumulate in the atmosphere, causing the rate of consumption of carbon dioxide by weathering reactions to increase. This increase would continue until the sink became equal to the source, after which atmospheric carbon dioxide would be in balance and no further change would occur. Evidently, the abundance of carbon dioxide in the atmosphere is determined by the requirement that the weathering sink be equal to the volcanic source.

Volcanoes result from high temperatures within the earth. There are several reasons for believing that internal temperatures were higher early in earth history than they are today. The most important is that gravitational energy converted into heat during the process of Earth's formation, both by the bombardment of the surface during accretion and by the settling to the center of dense nickel and iron at the time of internal differentiation. Early in earth history, therefore, the recently formed planet was still cooling down. In addition, there is a continuous source of heat within the earth that was larger in the past. This heat source is the decay of radioactive uranium, thorium, and potassium, present in small amounts in the minerals of which the earth is composed. Decay of these radioactive elements produces heat, but it also destroys the atom undergoing decay. The rate of radioactive heating has declined with the passage of time, by about a factor of 3, as the radioactive elements have gradually disappeared.

Higher internal temperatures probably resulted in more volcanic activity. If there were more volcanoes on the primitive earth

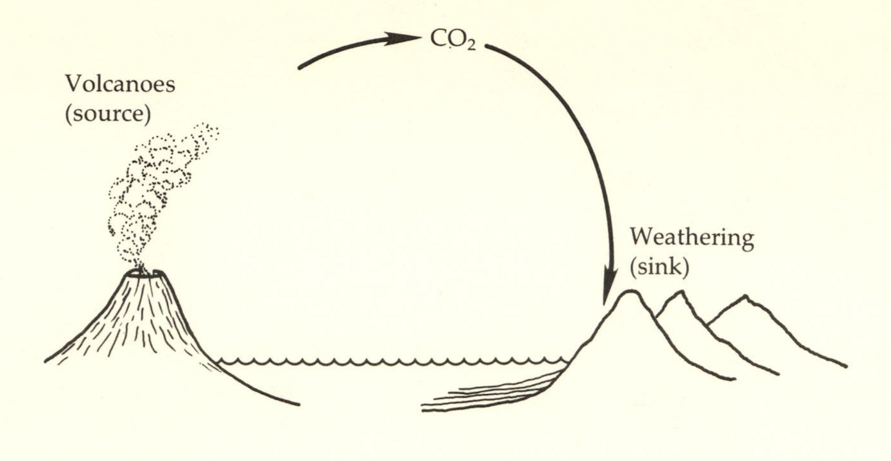

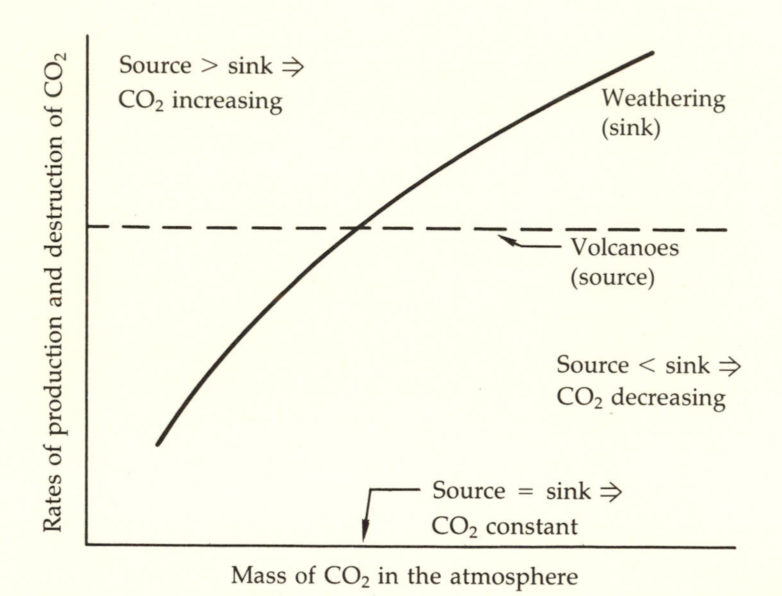

Figure 4–2. Carbon dioxide in the atmosphere. As atmospheric carbon dioxide increases, so does the rate of weathering. The source is independent of the amount of carbon dioxide already in the atmosphere.

than there are today, there was probably more carbon dioxide in the atmosphere. This conclusion can be made more quantitative by another argument, admittedly indirect. As the sun has aged it has converted light hydrogen nuclei into heavier helium nuclei by nuclear reactions, as described in Chapter 2. This increase in the average mass of the nuclei in the sun has resulted in increased pressures and higher temperatures in the solar interior. The higher temperatures have caused more rapid generation of energy by nuclear reactions within the sun and therefore a gradual increase in the brightness of the sun. Theoretical calculations indicate that the sun today is brighter by perhaps 30% than it was soon after its formation.

If solar luminosity was less early in the history of the solar system, we would expect that the earth was colder, because the sun maintains the temperature at the surface of a planet. In fact, it would have been so cold that all water on the surface of the earth would have been frozen if atmospheric composition had not changed with time. This important implication of astronomical theory was called to the attention of earth scientists in 1972 by C. Sagan and G. Mullen of Cornell University. Their calculations for constant atmospheric composition yielded a completely frozen earth until about 2 billion years ago. This theory is contradicted by the geologic record that shows sedimentary rocks deposited under water at all ages extending back to the time the rocks found at Isua, Greenland, were formed, 3.8 billion years ago.

The significance of atmospheric composition is that certain gases can raise the temperature of the ground by absorbing the heat radiation it emits and returning some of this radiation to the ground. This atmospheric influence on surface temperature is called the greenhouse effect. Today the greenhouse effect, caused mainly by water vapor and carbon dioxide, raises average surface temperature by about 40° C. The likeliest resolution of the apparent conflict between astronomy and geology that I have just described is that the greenhouse effect of the early atmosphere was larger than it is today.

There has been some debate about the gas responsible for this enhanced greenhouse effect. Sagan and Mullen favored ammonia, but because of its high solubility in sea water and its chemical instability in the presence of sunlight this gas now seems unlikely to have been sufficiently abundant in the atmosphere. A much more plausible alternative is carbon dioxide. Calculations by T. Owen and colleagues at the State University of New York at Stony Brook indicate that carbon dioxide together with water vapor could have provided the necessary greenhouse effect, provided the carbon dioxide pressure was a few hundred times its present value.

My own study of the balance between the rate of release of carbon dioxide by volcanoes and metamorphism and the rate of removal of carbon dioxide by weathering shows that such high pressures of carbon dioxide can reasonably be expected for the early earth, particularly when the temperature dependence of the rate of weathering is taken into account. Modern data on weathering rates under different climatic conditions show that they increase rapidly with increasing temperature. This dependence of weathering rate on temperature results in a negative feedback that tends to minimize excursions in the average surface temperature of the earth. Increasing solar luminosity results, at first, in an increase in surface temperature, but this leads in turn to more rapid weathering and therefore a decrease in the carbon dioxide pressure. A decrease in carbon dioxide diminishes the temperature increment of the greenhouse effect and Earth's surface cools down again. As solar luminosity has increased steadily over the age of the solar system, this feedback may have caused a marked decline in atmospheric carbon dioxide accompanied by a small increase in surface temperature (Figure 4–3).

Among the major atmospheric gases, nitrogen is quite different from carbon dioxide because it does not react with rocks. Most of Earth's nitrogen is in the atmosphere today, and the same would have been true early in earth history, once the initial release of volatiles from the interior was complete. The primitive atmosphere would therefore have contained about as much nitrogen as the modern.

Oxygen is almost the reverse of nitrogen. It does react with rocks and it is not released by volcanoes because it is so chemically reactive. Instead, it combines readily with elements such as iron that are abundant in the interior of the earth. Oxygen is present in Earth's atmosphere today almost entirely because of the activities of photosynthetic organisms. These organisms convert carbon dioxide into cell material and at the same time convert water into oxygen. Oxygen is absent in significant quantities from

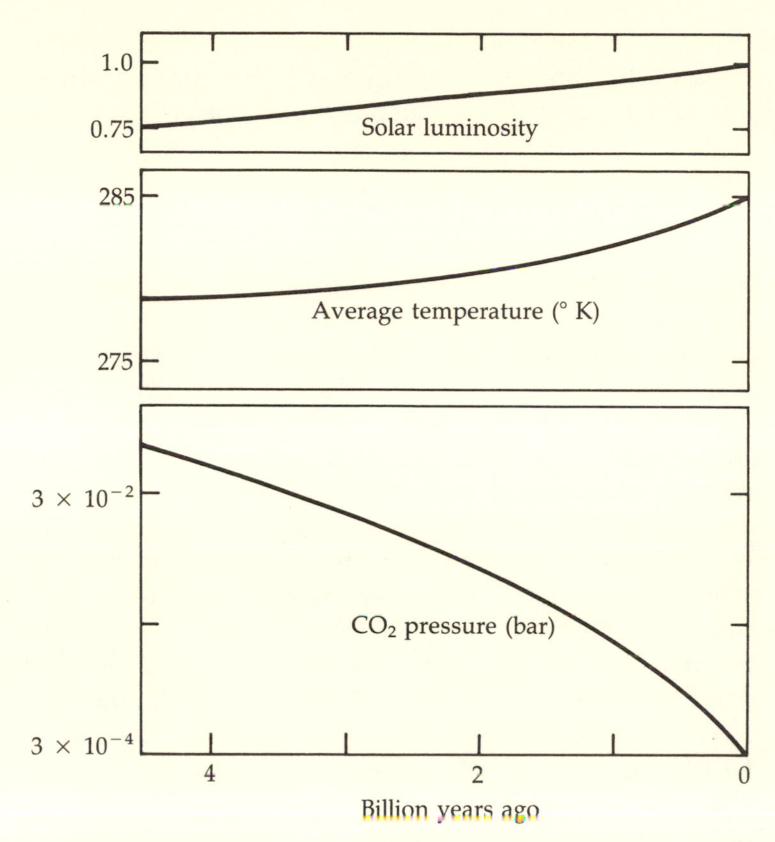

Figure 4–3. Surface temperature and atmospheric carbon dioxide in response to increasing solar luminosity. Solar luminosity is expressed on a scale relative to the present, where luminosity today equals 1. Carbon dioxide is expressed in bars, where 1 bar is the present total pressure at the ground.

the atmospheres of all of the other planets. Except for trace amounts produced by the action of sunlight on carbon dioxide and water vapor, it would have been absent also from the terrestrial atmosphere before the origin of life. Air on the early Earth was therefore anaerobic. Animals, if there had been any, could not have breathed.

Hydrogen is the inverse of oxygen. Because the two gases react readily, an increase in the concentration of one causes a decrease in the concentration of the other. The more abundant gas consumes the less abundant one in the chemical reaction that produces water (H_2O). There must have been more hydrogen in the anaerobic atmosphere than there is in the aerobic atmosphere of the modern world. How much more is not yet clear.

BIOGEOCHEMICAL CARBON CYCLES

My description of the control of carbon dioxide by a balance between its rate of release by volcanoes and metamorphism and its rate of removal by the weathering of silicate minerals and deposition of carbonate minerals is an adequate approximation only in an average sense over time scales of several million years or more. The picture is complicated over shorter periods of time by the activities of organisms, including mankind, and by temporary imbalances in the exchange of carbon dioxide between atmosphere and ocean.

As shown in Figure 4–4, the system can be analyzed in terms of reservoirs and the fluxes of carbon that link them. The atmospheric reservoir of carbon dioxide is closely coupled to two larger reservoirs, the surface organic reservoir composed of living and recently dead organisms, and the oceanic reservoir of bicarbonate ions in solution. Carbon dioxide is shared between atmosphere and ocean according to the laws of physical chemistry. The time required for the system to achieve a balance after a sudden change, say the addition of carbon dioxide to the atmosphere, is a few thousand years. Plants extract carbon dioxide from the atmosphere in the process of photosynthesis, converting it into surface organic matter. At the same time, the process of respiration converts organic carbon back into atmospheric carbon dioxide. This exchange achieves a balance in something like the time required for dead trees to decay, say a few hundred years.

Over longer time scales we are interested in the processes that either add or subtract carbon from the combined reservoirs of atmosphere, ocean, and surface organic matter. These are the processes of silicate weathering as a sink and volcanism and metamorphism as a source that I have described in the text. They are fast enough to cause significant change in the surface reservoirs in times of a million years or so. The dissolution of carbonate minerals exposed to rain water is neither a source nor a sink of carbon dioxide over the long term. Because the ocean is approximately saturated with calcium carbonate, dissolution of old carbonate minerals leads to deposition of an equal mass of new carbonate minerals.

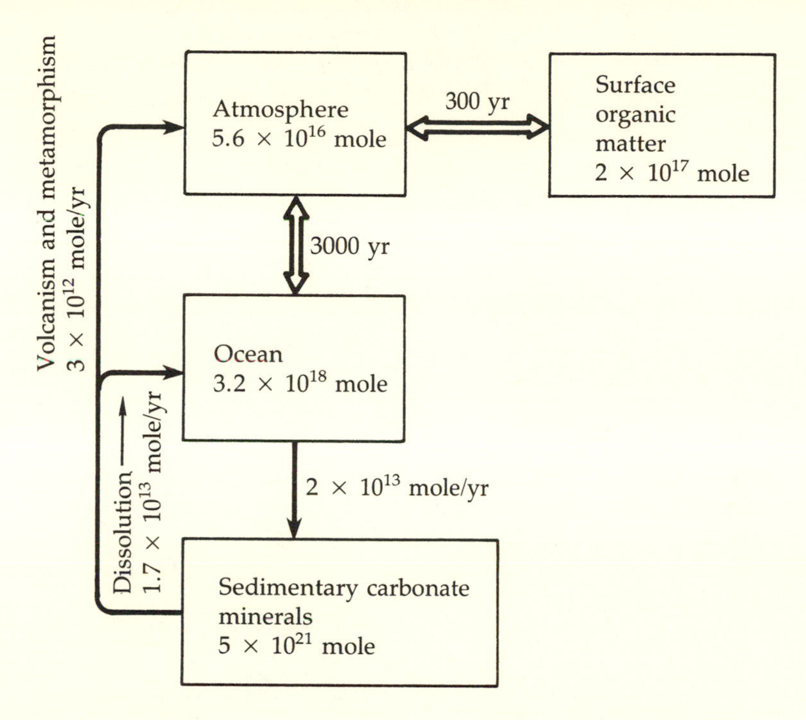

Figure 4–4. Reservoirs and fluxes of carbon. Reservoirs are expressed in moles (1 mole of carbon = 12 grams) and rates of transfer of material in moles yr^{-1}. The time scales for achievement of equilibrium between atmosphere, ocean, and surface organic matter are indicated.

Much of the scientific debate about the origin of life revolves around the question of whether or not the early atmosphere was highly reducing (rich in reduced gases like hydrogen, methane, and ammonia). Abiotic reactions (reactions that do not require the participation of living organisms) in highly reducing atmospheres are prolific sources of the organic molecules that were the building blocks of the earliest organisms. The abiotic synthesis of organic molecules in weakly reducing atmospheres (atmospheres of nitrogen, carbon dioxide, water, and small amounts of hydrogen) has so far proven to be much less fruitful.

After the initial period of degassing and the approach of atmospheric composition to a state of equilibrium, the hydrogen abundance would have depended on a balance between the volcanic source and a sink which is peculiar to hydrogen. Hydrogen is the lightest of all elements and thus is less tightly held by the force of earth's gravity than any other gas. There are always some hydrogen atoms at the top of the atmosphere traveling upwards fast enough to escape altogether and be lost in interplanetary space. On the primitive earth the volcanic source of hydrogen was balanced by this process of escape to space (Figure 4–5).

Detailed theories of the escape process have been developed and verified in the modern atmosphere. They show that the rate of escape is proportional to the concentration of hydrogen in the atmosphere with a constant of proportionality that is known within reasonable limits even for the primitive atmosphere. With this knowledge it is possible to calculate the hydrogen concentration in the primitive atmosphere for any assumed value of the volcanic source. If estimates of the present rate of release of volcanic hydrogen are used as a guide, the calculated hydrogen concentration is in the range 10^{-4} to 10^{-3} (one H_2 molecule per 10,000 to 1,000 air molecules). Contemporary estimates are most uncertain, and the rate of release of hydrogen may have been somewhat larger in the past. The balance between volcanic source and escape may possibly have yielded hydrogen concentrations as high as 10^{-2} (1%), but it is hard to imagine a larger concentration of hydrogen surviving for any geologically significant period of time.

My own view of the early atmosphere, shown in Figure 4–6, is that it was mainly nitrogen, like the modern atmosphere, with more carbon dioxide than there is today, no oxygen, and no more than 1% hydrogen.

The water vapor content of the atmosphere is governed by cycles of evaporation and precipitation and depends on the average surface temperature (Figure 4–7). There have been few climatic studies of ancient atmospheres, but it appears that the atmosphere I have described would have yielded surface temperatures somewhat lower than they are today. The early atmosphere may therefore have been drier. There is also a school of thought that favors a hot climate during the first few billion years of earth history. In my view, no adequate physical explanation has yet been offered of how a hot climate could have been sustained at a time when the sun was less bright than it is today.

This picture of the prebiological atmosphere, one much like that of the present day but weakly reducing, with no oxygen and small amounts of hydrogen, diverges strongly from the tradi-

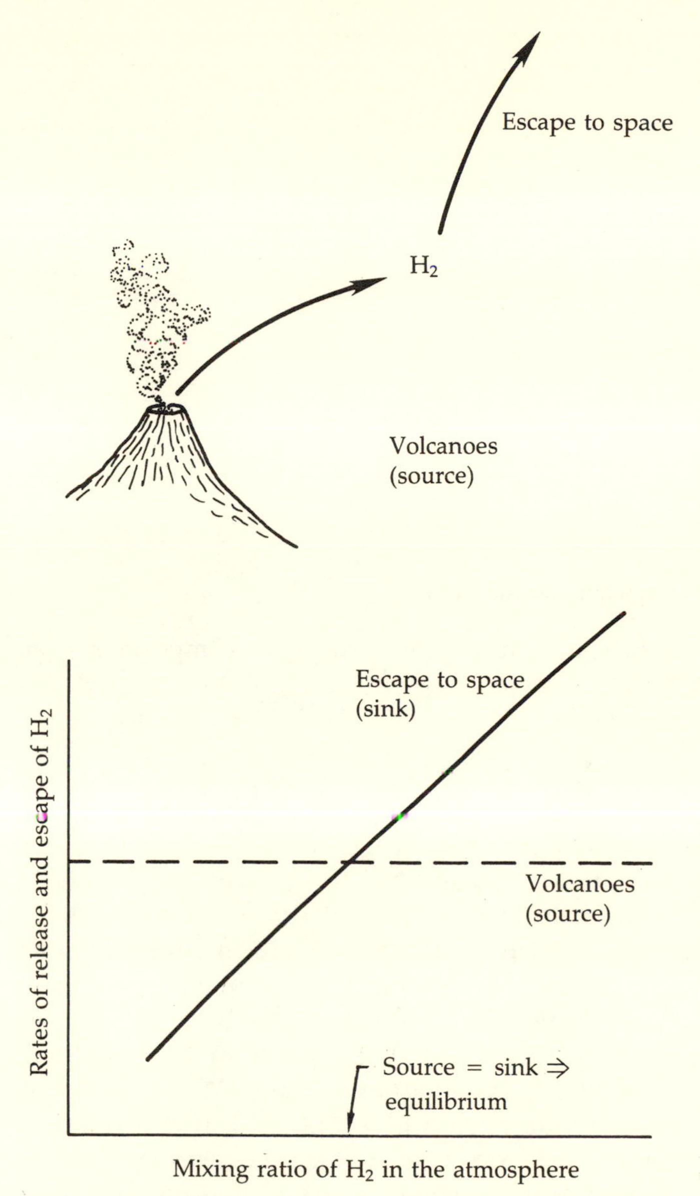

Figure 4–5. Hydrogen in the earth's early atmosphere.

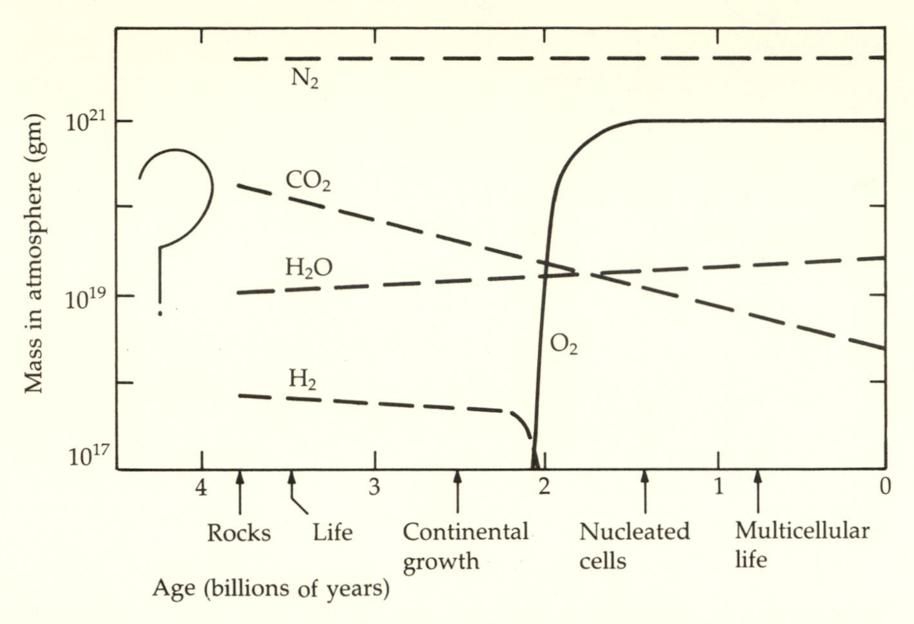

Figure 4–6. Schematic history of atmospheric composition since the beginning of the rock record. The dashed lines represent guesses. (Adapted from J. C. G. Walker, *Impact of Science on Society, 32,* 261, 1982.)

tional point of view, which favors a highly reducing atmosphere composed principally of hydrogen and compounds of hydrogen such as methane (CH_4) and ammonia (NH_3). A highly reducing atmosphere would have facilitated the origin of life because organic molecules form readily as a result of chemical reactions in such an atmosphere. I have not favored such an atmosphere because it could have been sustained only by a volcanic source of gas much more highly reduced (richer in hydrogen) than the gases released from any modern volcano. Which point of view is correct depends directly on how the earth was formed.

The degree of oxidation of volcanic gases (for our purposes the ratio of hydrogen to water vapor) is governed by the degree of oxidation of the rocks in the upper mantle, where the volcanic gases originate. The oxidation state of the upper mantle can be characterized by the ratio of metallic iron to iron oxide. The upper mantle has lacked metallic iron throughout observable earth history, so volcanic gases have been only weakly reducing. If the prebiological atmosphere was highly reducing, there must have

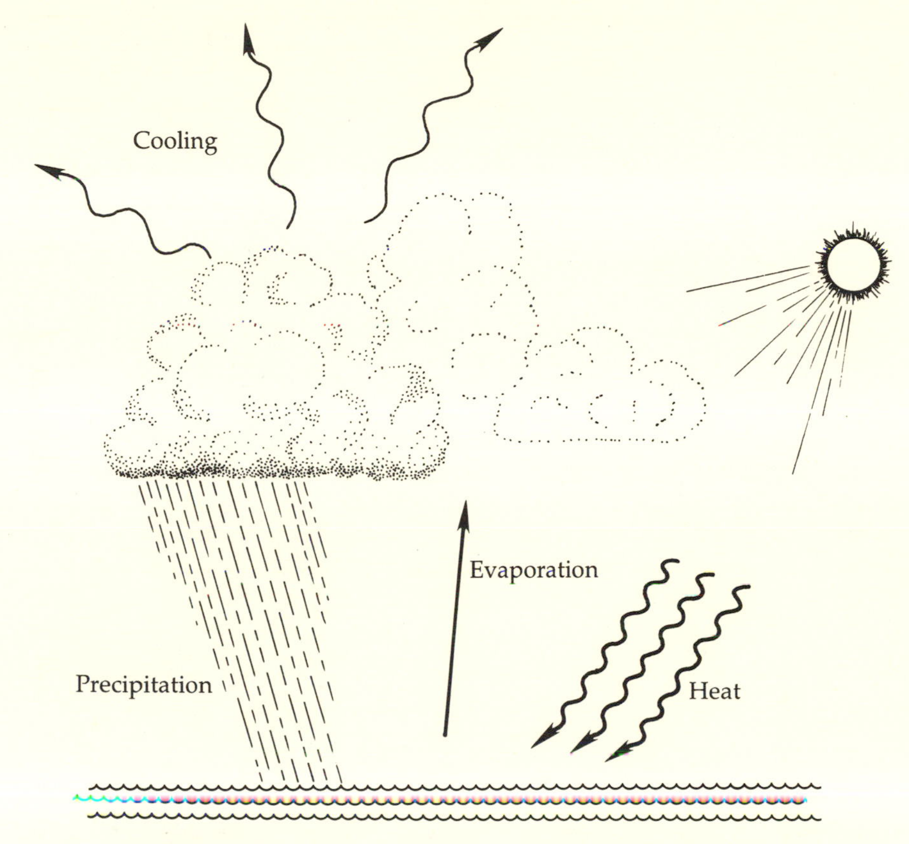

Figure 4–7. The effect of temperature on the water vapor content of the atmosphere.

been metallic iron in the upper mantle during the prebiological era, as suggested in Figure 4–8. Where is this metallic iron now? Presumably in the core. Thus the argument over whether life originated in a highly reducing or a weakly reducing atmosphere comes down to an argument over the relative timing of core formation and the origin of life. My own view inclines toward formation of the core during the process of growth of the planet. Perhaps life originated while Earth was still accreting or very shortly thereafter, while a recently degassed atmosphere was still adjusting to its equilibrium composition. If not, then my guess is that life originated under a weakly reducing atmosphere.

"Guess" is not an unduly cautious word to use in this connection. We cannot yet discuss the early atmosphere and ocean

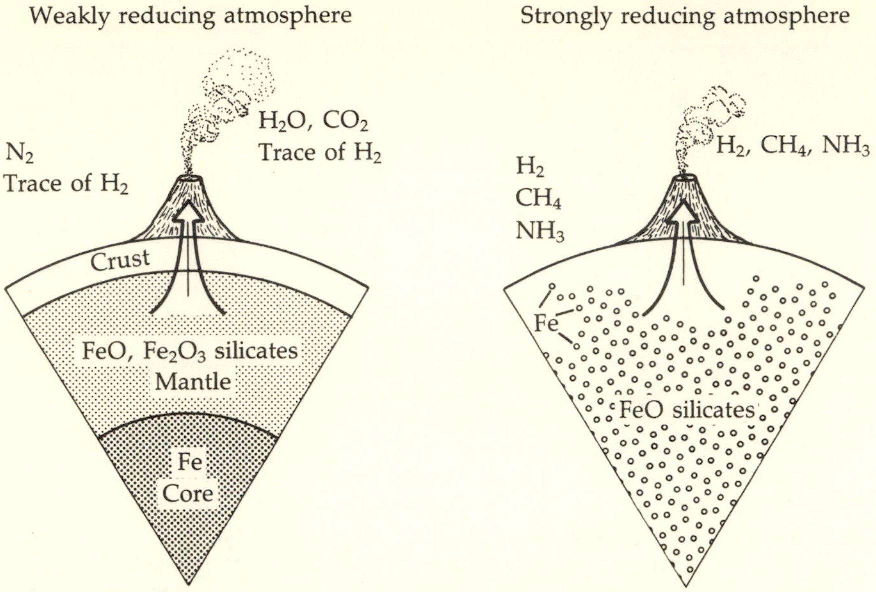

Figure 4–8. Contrasting views of the prebiological atmosphere.

OXIDATION AND REDUCTION

Reactions involving oxidation and reduction will be mentioned frequently in this book. The feature that defines an oxidation–reduction reaction is the transfer of electrical charge from the element being oxidized to the element being reduced. The oxidation of iron provides an illustration.

In pyrite (FeS_2), *the iron is in its* $+2$ *valence state; that is, it lacks two negatively charged electrons. The sulfur is in its* -1 *valence state; it has one electron too many. The compound,* FeS_2, *is electrically neutral. If pyrite is oxidized to hematite and sulfur dioxide*

$$4FeS_2 + 11O_2 \rightarrow 2Fe_2O_3 + 8SO_2$$

oxygen is consumed while the valence of the iron changes to $+3$ *and that of the sulfur changes to* $+4$. *Oxidation therefore corresponds to an increase in the positive charge on the element being oxidized. The ferric iron* (Fe^{+3}) *in hematite is more highly oxidized than the ferrous iron* (Fe^{+2}) *in pyrite.*

Oxygen is not necessarily involved in an oxidation reaction. What is needed is an element (the electron acceptor) to accept the electrons being lost by the element undergoing oxidation (the electron donor). Oxygen serves as electron acceptor in the reaction cited above; its valence changes from 0 *to* −2. *Hydrogen can also serve as electron donor. In the reaction*

$$Fe_2O_3 + H_2 \rightarrow 2FeO + H_2O$$

for example, as the hydrogen is oxidized to water, its valence changes from 0 *to* +1. *The iron is reduced, by loss of oxygen, from a valence state of* +3 *to one of* +2. *The hydrogen is therefore called a reducing gas.*

with much conviction. There are few constraints on speculation. In this chapter I have tried to show the kinds of arguments and evidence that illuminate the subject, but it will be some time before a scientific consensus emerges.

Suggested Reading

Holland, H. D. *The Chemical Evolution of the Atmosphere and Oceans.* Princeton, NJ: Princeton University Press, 1984.

Schopf, T. J. M. *Paleoceanography.* Cambridge, MA: Harvard University Press, 1980.

Walker, J. C. G. *Evolution of the Atmosphere.* New York: Macmillan, 1977.

CHAPTER 5

First Life

WHAT ARE the essential attributes of the simplest organism that is unmistakably alive? First, it can grow by synthesizing the material of which it is composed out of nutrients taken in from its surroundings, a process known as biosynthesis. Second, because the chemical reactions that take place in biosynthesis require a supply of energy, the simplest living creature must be able to generate energy. The combined processes of energy generation and biosynthesis are called metabolism; thus, third, a living organism has a metabolic apparatus. Fourth, it also has a genetic apparatus to control the biosynthetic machinery, ensuring that the right materials are produced. And fifth, for life to endure, organisms must reproduce, passing genetic information on to their offspring.

Simple organisms with little more than these basic attributes exist today. They are the bacteria, microscopic organisms consisting of a single cell, a blob of protoplasm surrounded by a cell wall. The cell contains metabolic and genetic machinery that causes the organism to grow under favorable conditions. The growing cell divides into two essentially identical offspring cells, and these cells grow and divide again. Although it is possible to speculate about forms of protolife that were less developed, the bacteria provide a reasonable guide to the nature of the earliest organisms. The fossilized remains of bacteria that lived 3.5 billion years ago have recently been found in the Warrawoona rocks near North Pole in Western Australia.

The metabolic activity in bacteria and all other organisms is coordinated and carried out by proteins. Proteins called enzymes catalyze the many reactions required for biosynthesis and for energy production. As catalysts, they cause chemical reactions to occur between other compounds and they control the rates of these reactions. Other proteins serve as structural elements.

A protein is a large molecule consisting of a long chain, called a polymer, of amino acids chemically bonded one to the next. The amino acids are compounds of some tens of atoms, carbon, hydrogen, oxygen, nitrogen, and sulfur. Only 20 amino acids are found in proteins. Nonetheless, an enormous variety of proteins results from combining these 20 building blocks (called monomers) in different sequences in the polymer. The sequence of amino acids controls the way in which the polymer coils and folds upon itself to produce a large molecule with a specific shape and chemical activity. Protein molecules typically contain several hundred monomers (amino acid molecules).

The synthesis of proteins within the cell is controlled by the genetic apparatus, the key component of which is deoxyribonucleic acid (DNA). DNA is another polymer, a double-stranded chain with many thousands of links made up of only four different monomers called deoxyribonucleotides. The sequence of these monomers in DNA constitutes a code that determines the order in which amino acids are added to a growing protein polymer during the synthesis of new proteins. The deoxyribonucleotides are compounds of a sugar molecule (called 2-deoxyribose) with simple organic molecules called purine and pyrimidine bases. What gives DNA its unique genetic role is the fact that it can be used within the cell to synthesize a second chain of DNA with an identical sequence of purine and pyrimidine bases. Thus, the genetic information required for the synthesis of proteins can be reproduced and shared with offspring cells.

Essential aspects of the origin of life, then, are the origin of metabolic machinery, dependent on proteins, and the origin of genetic machinery, dependent on DNA. How were these complex macromolecules first formed? How did their complementary activities achieve the necessary coordination? The answers are not yet known, so I can give only a cursory account of current ideas.

Chemical evolution is the name given to the processes that made possible the appearance of life on our previously lifeless

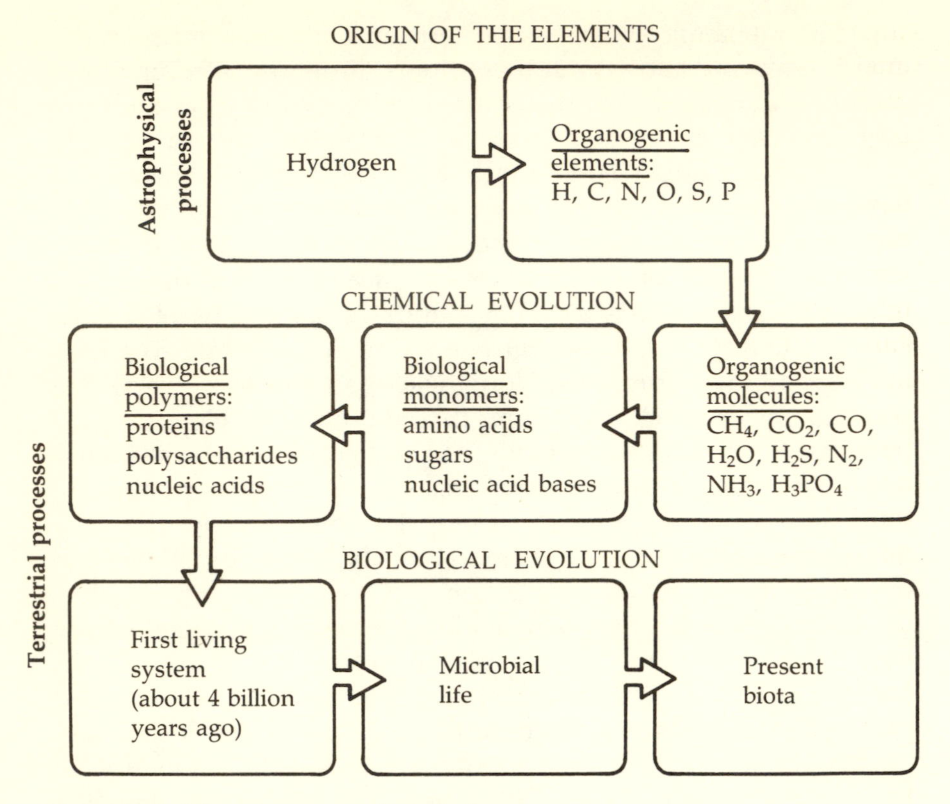

Figure 5–1. The sequence of evolution.

planet (Figure 5–1). During the course of chemical evolution, increasingly elaborate chemical compounds, the components of life, were gradually built out of the simple inorganic molecules of the prebiological atmosphere and ocean.

Chemical evolution is believed to have started with the abiotic synthesis of simple organic molecules such as hydrogen cyanide (HCN) and formaldehyde (H_2CO). Possible mechanisms for the formation of these molecules are known, although there is uncertainty about their rate of production. What is needed is a source of energy to break the unreactive atmospheric molecules into reactive constituents that can combine with one another to form new molecules. Lightning and ultraviolet radiation from the sun would have been suitable energy sources on the early earth.

Mechanisms are also known by which these simple organic molecules could have reacted to produce essential biological building blocks (biological monomers) such as amino acids, sugars, and purine and pyrimidine bases. Many laboratory experiments have demonstrated that these building blocks are produced readily when gaseous mixtures that might have characterized the primitive atmosphere are exposed to a source of energy.

Organic molecules formed in the primitive atmosphere could have settled to the surface or dissolved in the ocean. Any that remained in the atmosphere would have been disrupted by the very energy sources that initiated their synthesis. Even in the relatively sheltered environment of the ocean they would have suffered gradual destruction by reactions with other dissolved molecules or by spontaneous thermal disruption. In time, the concentrations of organic constituents would have come to depend on a balance between the rate of synthesis and the rate of destruction. Stable compounds would have been more abundant than unstable ones. Essential to the preservation of organic compounds for any significant period of time was the absence of oxygen. Organic molecules are reducing; most of them react readily with free oxygen to produce simpler organic compounds as well as carbon dioxide and water. Because most of the destruction reactions occur at rates that increase rapidly with temperature, preservation would have been favored by a cold Earth.

The next stage of chemical evolution was the assembly of biological polymers out of large numbers of the individual building blocks. The formation of polymers under primitive earth conditions has been much harder to demonstrate in the laboratory than the formation of amino acids and similar monomers. It seems clear that the balance described above between synthesis and destruction of organic molecules could not have provided concentrations in the open ocean that were high enough for polymer formation. Mechanisms to concentrate the building blocks were required. Possible concentration mechanisms include evaporation of water in isolated basins, freezing (which leaves dissolved matter in solution), and a process of adsorption in which certain compounds in solution are selectively removed by becoming attached to the surface of a solid. The mineral constituents of common clay have proved to be effective agents for the concentration of organic molecules by adsorption. It has been suggested that the regular crystal structures of these minerals may have imposed some order on the sequence in which concentrated monomers were connected to form organic polymers.

Somehow, solutions of organic compounds were concentrated, and polymers were formed. Some of these polymers resembled proteins and functioned like enzymes. (Enzymes catalyze particular biochemical reactions, increasing the rates of reaction without being consumed in the process.) Enzymes on the primitive earth would have altered the composition of the primordial organic soup by accelerating the rates at which some organic compounds were made while enhancing the rates of destruction of other compounds.

The formation of polymers from their component monomers typically requires an input of energy. Conversely, energy is released when polymers and other organic molecules degrade into simpler compounds. In some unspecified way the processes of degradation and biosynthesis became coupled (Figure 5–2) to yield a metabolic system in which the energy released by degradation of some organic compounds was used to synthesize new and more complex molecules. Initially this metabolic activity may have proceeded without the direction of a genetic apparatus, but apparently some of the polymers formed were nucleic acids, capable of controlling the synthesis of proteins. In some unspecified way the genetic mechanism became coupled to the metabolic mechanism to provide it with direction and continuity as well as

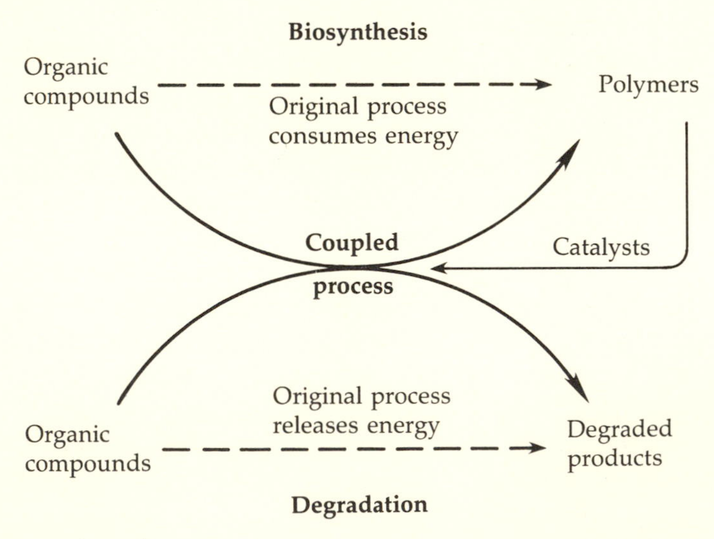

Figure 5–2. Coupling degradation (energy production) and biosynthesis (energy consumption). (After S. Black, *Perspectives in Biology and Medicine, 21*, 348, 1978.)

the opportunity for evolution by natural selection (or biological evolution).

In addition to metabolic and genetic machinery, all organisms contain membranes made out of proteins and other kinds of polymers. Some membranes enclose the cell, separating living protoplasm from the surrounding medium, and others are internal membranes. Recent research has shown that membranes are not simply containers; many of the biochemical processes of the cell take place on membranes. An active function of the cell wall, for example, involves the extraction from the surrounding medium of constituents needed by the cell and the excretion of waste products to the surroundings. Chemical evolution probably had to produce at least crude biological membranes before life could originate.

The appearance of cells, complete with membranes and walls, metabolic machinery, and genetic apparatus, opened the way for biological evolution (Figure 5–1). The processes of mutation and natural selection, the causes of biological evolution, operate on populations of cells or populations of organisms made up of cells. The chemical era of earth history ended with the origin of the cell. Unfortunately, chemical evolution has left no unambiguous traces in the geological record. Our understanding of the origin of life is based on theory, laboratory simulation, and analogy.

Suggested Reading

Margulis, L. *Early Life*. Boston: Science Books International, Inc. (now Jones and Bartlett Publishers, Inc.), 1982.

Miller, S. L., and L. E. Orgel. *The Origins of Life on Earth*. Englewood Cliffs, NJ: Prentice-Hall, Inc., 1974.

PART TWO *Interactions*

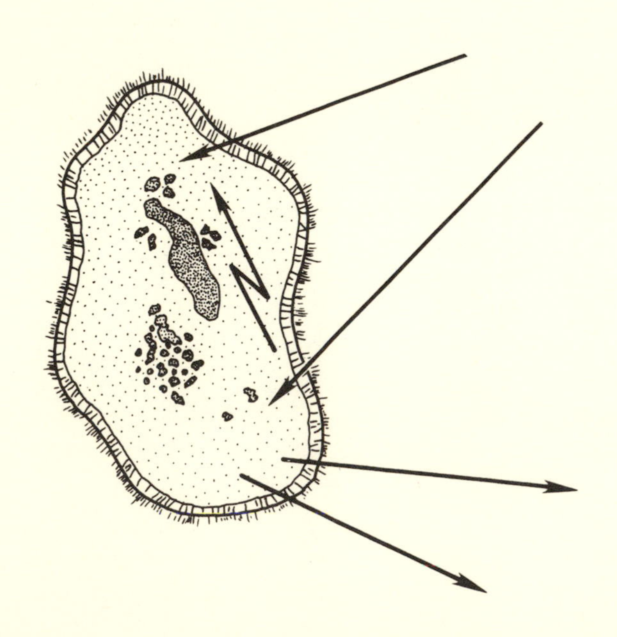

CHAPTER 6

The Age of Microbes

THE EARLY STAGES of life on Earth were characterized by the coordinated evolution of the metabolic capabilities of microbes and the composition of the atmosphere. The development of new metabolic capabilities affected the composition of the atmosphere and the resultant changes in the atmosphere stimulated the evolution of still new metabolic capabilities. Again, there is no clear geological evidence for the story I am about to tell; it is based entirely on speculation. It is also controversial in key elements, as I shall note in what follows.

The origin of life was preceded by the abiotic synthesis of organic molecules in the course of photochemical reactions in the primitive atmosphere. The first organisms presumably exploited this supply of organic material in their environment. They were therefore heterotrophs. They built themselves out of whatever organic molecules they found around them, and derived energy for their very limited biological activity from the chemical reactions of these same organic molecules, as suggested in Figure 6–1. (Organisms that derive energy from chemical reactions are called chemotrophs. The words can be combined. Thus, chemoheterotrophs require an external source of organic compounds and exploit chemical energy.)

Fermentation is the metabolic process by which an organism obtains energy by oxidizing one organic compound with another. It is the most widespread of all the metabolic processes that yield energy, as well as the one with the least complex biochemistry.

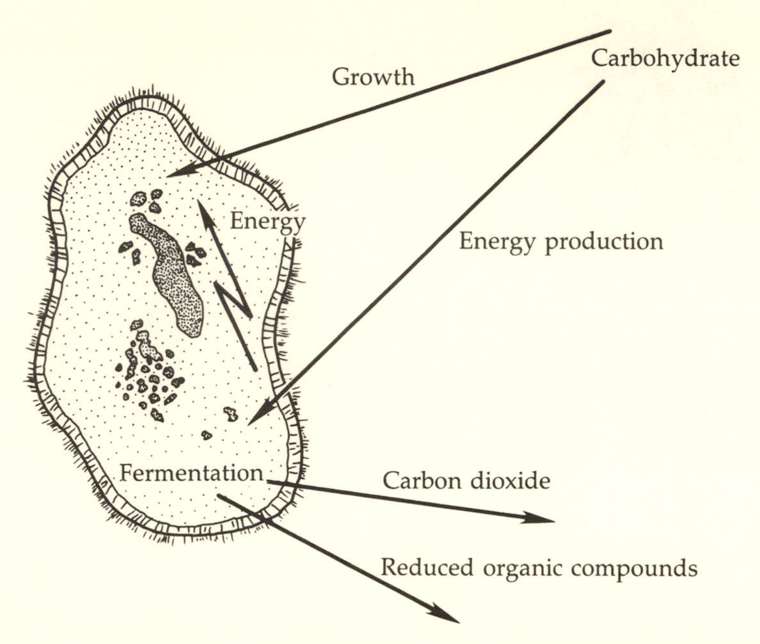

Figure 6–1. A primitive microbe deriving energy from the fermentation of organic compounds in its environment and using the energy for growth by incorporating available compounds into its cellular material. (Adapted from J. C. G. Walker, *Origins of Life, 10,* 93–104, 1980.)

EVOLUTION

The different meanings of this word can cause confusion in a discussion of earth history. The word by itself, without qualification, implies little more than progressive change with time. Thus, the sun evolves as it consumes its nuclear fuel, atmospheric composition evolves as different processes release or remove atmospheric gases, and my taste in clothes evolves as I grow older.

Similarly, biological evolution need not mean more than the change with time in the kinds of organisms inhabiting the earth. Frequently, however, it is taken to imply a specific mechanism of change, generally the change outlined in Darwin's theory of the origin of species by means of natural selection (which will be discussed in Chapter 12).

Biological evolution is clearly a property of life, but chemical evolution is a presumed property of the world before life. The term "chemical evolution" describes the physical and chemical processes that are thought to have preceded biological evolution and to have brought about the conditions that permitted life to arise.

On both counts, therefore, it is a strong candidate to have been the first metabolic source of energy. There is, however, no general agreement that the earliest organisms gained energy by fermentation of abiotically synthesized organic molecules. Disagreement exists because of uncertainty about whether abiotic synthesis could have produced organic compounds in quantities great enough to sustain life. The problem is particularly severe in the kind of weakly reducing atmosphere that I have advocated in Chapter 4. It has been argued that the first organisms were not heterotrophs but photoautotrophs, able to capture the energy of sunlight, rather than chemical energy, to synthesize cell material directly from carbon dioxide instead of from abiotically synthesized organic compounds. I shall set this possibility aside for now while exploring the consequences of a heterotrophic origin for life.

Two factors prevented biological evolution from stagnating after the origin of heterotrophy. One is that microbes in a favorable environment grow and multiply until they have exhausted the supply of some essential nutrient. This is called Liebig's Law, and is the equivalent for microbes of the Malthusian hypothesis concerning human populations. The other is that the genetic apparatus does not always produce identical offspring. Variability is introduced among higher organisms by combining genetic information from two parents in a single offspring (Figure 6–2). Prokaryotes, by contrast, reproduce by simple division of a single cell, so change occurs only when an error occurs in the preservation and duplication of genetic material. Mutation is the name given to the process of genetic change that produces cells with new properties. Many, but not all, mutations are harmful because they result in offspring that are less able than their ancestors to survive in a competitive world. Some mutations are useful. Among primitive microbes, mutation from time to time produced cells with metabolic capabilities that had previously been lacking. If the new metabolic capabilities gave their possessors an advantage in the competition for resources, the possessors were able to grow and multiply until the new metabolic capability became commonplace. Mutation combined with natural selection therefore caused microbes to evolve. The direction of evolution depended on environmental factors and on the nature of the competition with other organisms, both of which distinguished beneficial mutations from harmful ones.

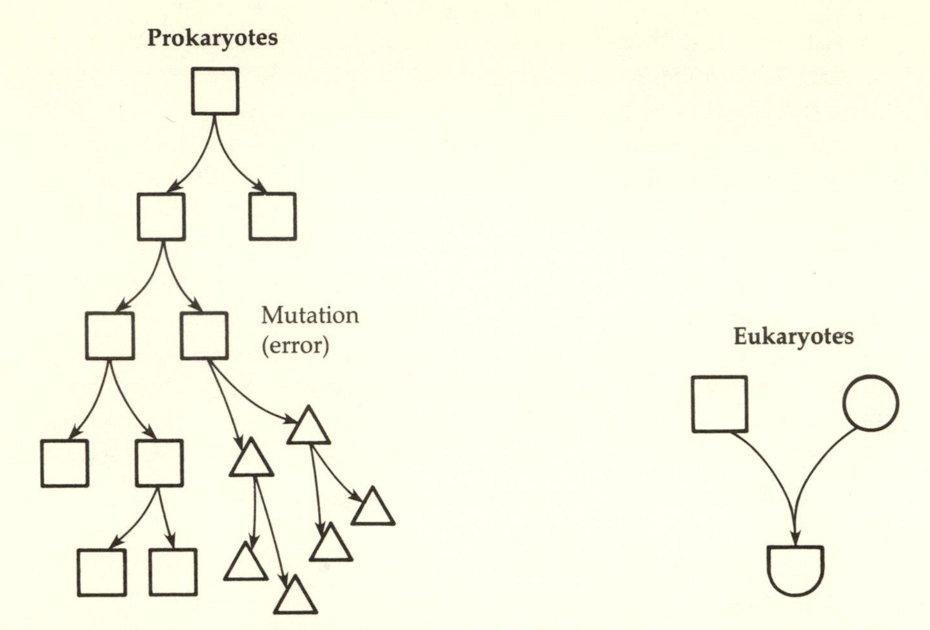

Figure 6–2. Reproduction of eukaryotic and prokaryotic organisms. Sexual reproduction among eukaryotic organisms produces offspring inheriting features of both parents. Prokaryotic organisms (which reproduce by division of a single cell) have offspring that are genetically identical to the parent unless an error occurs.

FERMENTATION

In a typical fermentation reaction, carbon serves as both electron donor and electron acceptor. An example is the fermentation that converts sugar to alcohol and carbon dioxide:

$$C_6H_{12}O_6 \rightarrow 2C_2H_5OH + 2CO_2 + \text{energy}$$

The valence of the carbon changes from 0 in the sugar to −2 in the alcohol and +4 in the carbon dioxide. Some of the sugar is therefore oxidized to carbon dioxide and the rest is reduced to alcohol in the process, while energy is released.

Alcoholic fermentation is only one of many fermentation reactions that occur in nature, and it is most unlikely that it was the first. Many organic compounds besides sugar are subject to fermentation. Unlike alcoholic fermentation, many fermentation reactions involve different organic compounds as electron donor and electron acceptor.

At first the population of microbes was small and the supply of organic molecules was abundant and diverse. The first organisms found the compounds they wanted readily available in their surroundings. Gradually, however, microbial life exhausted the supply of one essential nutrient after another. The total mass of organic material involved was never more than a very small fraction of the mass of the atmosphere. Further expansion of life was not possible when one of these essential compounds was absent from the environment until some organism had acquired, by mutation, the ability to synthesize the needed compound from other more abundant materials. In this way, mutation and natural selection led to the step-by-step enhancement of the biosynthetic capabilities of microbes while the microbes themselves removed one organic compound after another from their surroundings.

In due course, however, expanding life encountered a problem that could not be overcome by the introduction of a new biosynthetic capability. Biosynthesis, like the synthesis of plastic, dissipates energy, and life exhausted its supply of readily available energy. Fermentation converts molecules of intermediate levels of oxidation into compounds from which energy cannot be extracted by fermentation; their chemical energy content has been used up. Eventually, organic molecules were consumed in fermentation as fast as they were produced by abiotic synthesis. For a time thereafter, the expansion of life was limited by the abiotic source of organic molecules (Figure 6–3).

Evolution released life from this constraint by producing an organism able to tap a new source of energy. There is no general agreement on what this new source was, but I favor the theory that it was chemical energy represented by the mixture of reduced and oxidized gases in the primitive atmosphere. There are microbes in existence today, called methanogenic bacteria, that derive energy from a reaction between hydrogen and carbon dioxide to produce methane (CH_4) and water. They use the energy released by this reaction to synthesize cell material from inorganic nutrients. Because hydrogen and carbon dioxide were both present in the primitive atmosphere, I think the first autotroph (actually a chemoautotroph because it exploited a chemical source of energy) may have resembled the methanogenic bacteria.

The autotrophs would have flourished initially as they converted abundant supplies of carbon dioxide and hydrogen into methane and water. Photochemical reactions in the atmosphere

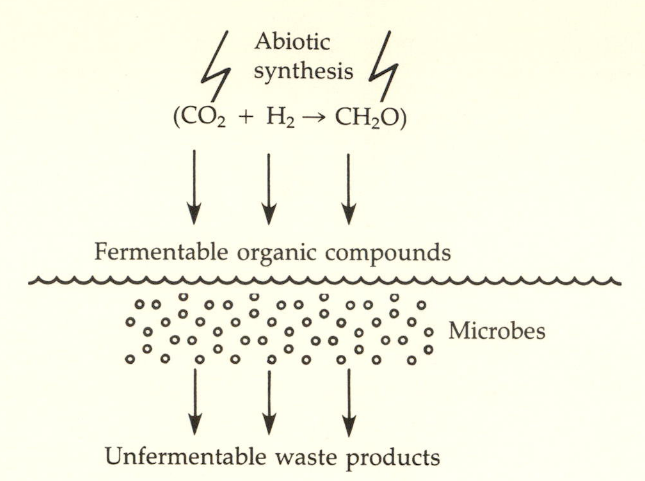

Figure 6–3. The expansion of life was limited by the rate of supply of fermentable organic compounds. (Adapted from J. C. G. Walker, *Origins of Life, 10,* 93–104, 1980.)

converted the methane back into hydrogen and carbon dioxide. The heterotrophs would have flourished also, enjoying the abundance of food synthesized by the autotrophs. This situation could not long endure, however, because the heterotrophs produced an unfermentable organic residue that sank to the bottom of the sea and disappeared into sediments. The combined efforts of autotrophs and heterotrophs therefore extracted hydrogen and carbon dioxide from the atmosphere and converted them into solid organic compounds in sediments.

This was the beginning of a significant geochemical role for living organisms, as illustrated in Figure 6–4. I believe that the first autotrophs reduced the concentrations of hydrogen and carbon dioxide in the atmosphere. These gases came from volcanoes (see Chapter 4), and because the source of carbon dioxide was larger than the source of hydrogen, the supply of hydrogen would have been exhausted first. In accordance with Liebig's Law, the autotrophs presumably lowered the amount of hydrogen in the atmosphere to a level at which they could barely survive. (Did they begin to suffocate?) Life could not expand any faster than permitted by the supply of fresh hydrogen from volcanoes.

Life evolved ways to overcome this second energy crisis with

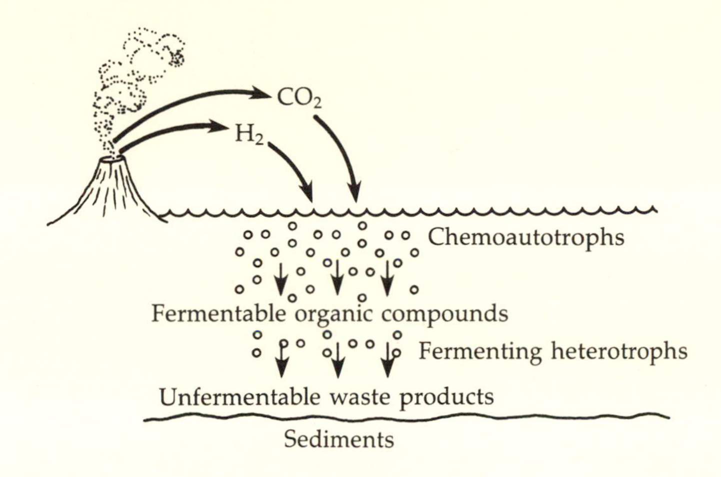

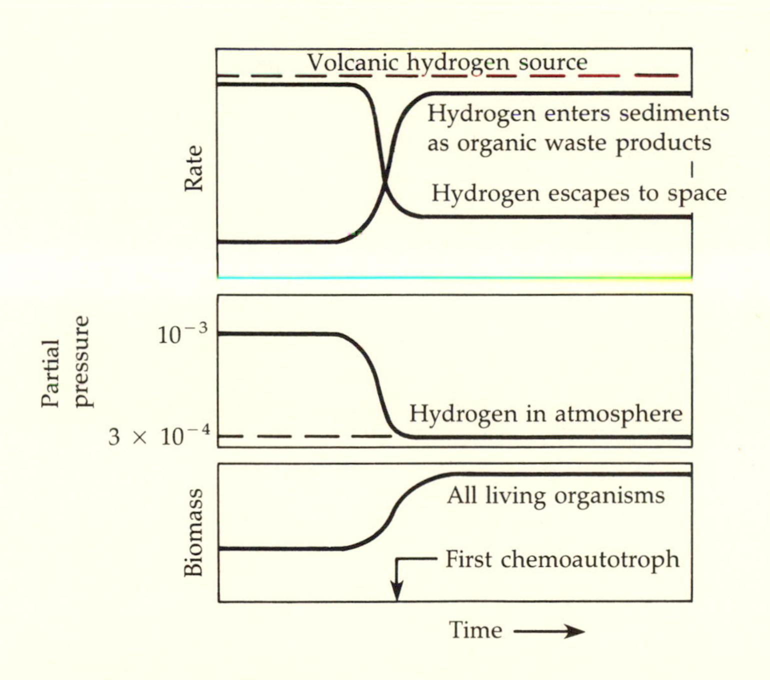

Figure 6–4. After the appearance of the first chemoautotrophs, biomass increased. The combined activities of chemoautotrophs and fermenting heterotrophs decreased the hydrogen content of the atmosphere, converting hydrogen that previously had escaped to space into organic waste products incorporated in sediments. (Adapted from J. C. G. Walker, *Origins of Life, 10*, 93–104, 1980.)

its customary ingenuity. There emerged an organism possessing pigments (a colored organism), able to absorb the energy of sunlight and to use this energy to synthesize organic molecules. Photosynthesis, as this process is called, typically involves changing the carbon in carbon dioxide to a more reduced form, such as that in carbohydrate. As described in Chapter 4, an electron donor species must provide the electrons to change the valence of the carbon. In the familiar kind of photosynthesis carried out by plants and algae today, the electron donor is water. As carbon dioxide is reduced to carbohydrate, water is oxidized to free oxygen, which is released to the environment as a waste product. But the direct utilization of light energy by organisms almost certainly started with a simpler process called bacterial photosynthesis (Figure 6–5). Bacterial photosynthesis is possible only if a more reduced compound is available to donate electrons. Examples of suitable reactants are hydrogen (H_2) and hydrogen sulfide (H_2S), as well as organic compounds. Bacterial photosynthesis does not result in the release of oxygen to the environment. Its waste products depend on the reactant. They would be water if the reactant is hydrogen and sulphuric acid if the reactant is hydrogen sulfide.

By exploiting an abundant supply of solar energy, bacterial photosynthesis permitted a higher level of biological activity to be sustained by a given supply of reduced gases from volcanoes. The volcanic supply still limited the expansion of life, however, because some organic matter escaped to the bottom of the sea and became a part of the sediments. This drain on biospheric reduced compounds persists even in the aerobic world of today. Because both bacterial photosynthesis and fermentation require a supply of reduced compounds, life could not expand any faster than this supply was replenished by volcanoes (Figure 6–6).

Photosynthetic bacteria, which exploited the energy of sunlight and used hydrogen simply as a source of electrons, were able to flourish at lower hydrogen concentrations than their predecessors, the chemoautotrophs, which depended on the chemical reaction between hydrogen and carbon dioxide as their source of energy. Because the photosynthetic bacteria would have grown and multiplied until hydrogen was in short supply, they presumably depressed atmospheric hydrogen to levels even lower than previously.

To sum up, we can say that the first stages of the microbial

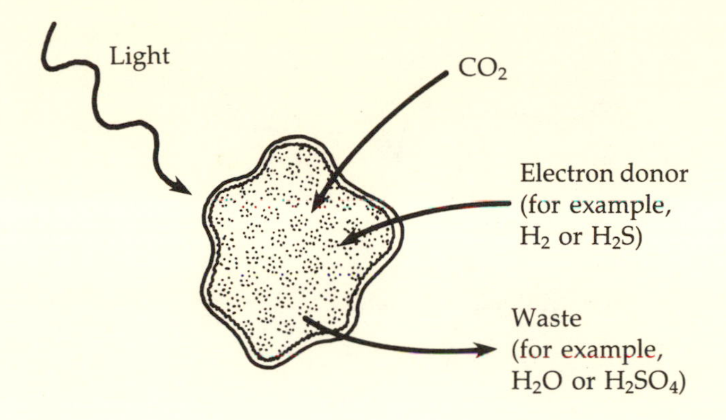

Electron acceptor		*Electron donor*	*Light*		*Waste*
CO_2	+	$2H_2$	$\rightarrow$	CH_2O +	H_2O
$2CO_2$	+	$H_2S + H_2O$	$\rightarrow$	CH_2O +	H_2SO_4

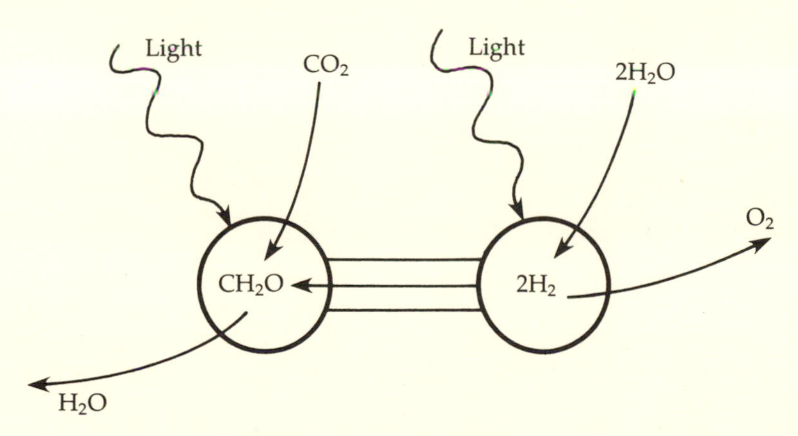

Figure 6–5. Some reactions of bacterial photosynthesis (top). Oxygenic photosynthesis involves two photochemical reaction centers (bottom). The first produces a reductant that reduces carbon dioxide to organic compounds as in bacterial photosynthesis. The second oxidizes water to oxygen. In these figures, CH_2O indicates organic matter and H_2 indicates reducing power.

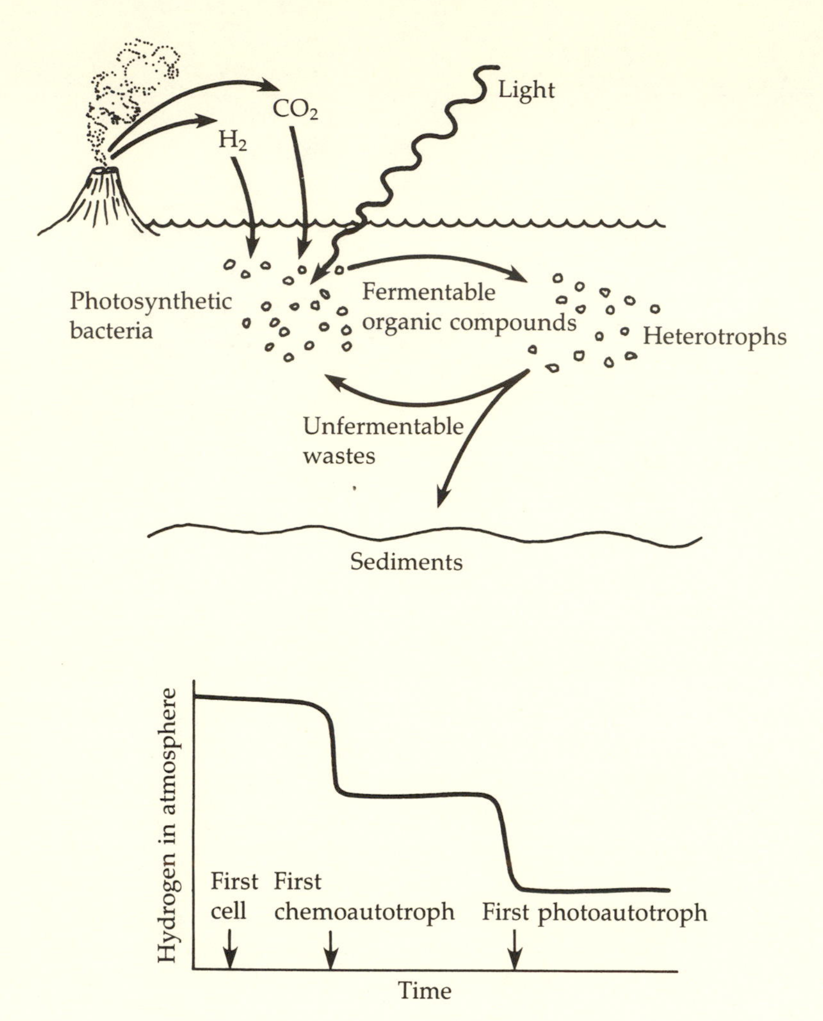

Figure 6–6. Cycling organic matter. Photoautotrophy caused another reduction in atmospheric hydrogen, but permitted the first rapid biological cycle of organic carbon. (Adapted from J. C. G. Walker, *Origins of Life, 10*, 93–104, 1980.)

era of earth history involved a close relationship between metabolism and the abundance of hydrogen. If the first organisms were fermenting heterotrophs, their expansion would have been limited by the rate of abiotic synthesis of fermentable organic compounds. The atmospheric mixture of carbon dioxide and hydrogen (and other reduced gases) represented a resource that primitive organisms could use to overcome this limitation. Once they had learned how, they exploited this resource until the hydrogen was exhausted. First, they depleted hydrogen to the point where further growth based on the chemical energy released in the reaction between hydrogen and carbon dioxide was not possible. Then some organisms evolved that used the energy of sunlight in the process of bacterial photosynthesis, oxidizing hydrogen to water and reducing carbon dioxide to organic material, a process that depressed atmospheric hydrogen even further.

It is possible, however, that life began with photoautotrophs. In this case, the first stages of this hypothetical sequence involving the joint evolution of atmosphere and metabolism would have been bypassed. Fermentative heterotrophy and chemoautotrophy would have originated later to exploit the waste products of the photoautotrophs. The photoautotrophs would have expanded, in accordance with Liebig's Law, until they exhausted the atmospheric supply of reduced reactants. In either case, after the origin of bacterial photosynthesis (photoautotrophy), the volcanic source of reduced gas was inadequate to offset the steady drain of reduced material as organic compounds incorporated into sediments. If life was to expand further it had to eliminate its dependence on reduced compounds as reactants in photosynthesis.

Life broke the "hydrogen habit" by learning how to dump unwanted oxygen into its surroundings. This considerable biochemical achievement and its geochemical consequences are discussed in the next chapter.

SUNBURN ON THE EARLY EARTH

Phototrophy introduced a new and potentially serious environmental problem with which life had to contend. The primitive atmosphere contained no oxygen and therefore no ozone, a molecule consisting of three atoms of oxygen. Ozone in the modern atmosphere shields the ground from biologically harmful solar ultraviolet radiation. The primitive atmosphere may have lacked such a shield. The presence of ultraviolet light at the earth's surface did not matter to the earliest organisms because they could live buried in mud or under rocks. Indeed, ultraviolet radiation was beneficial in that it initiated the photochemical reactions that led ultimately to the

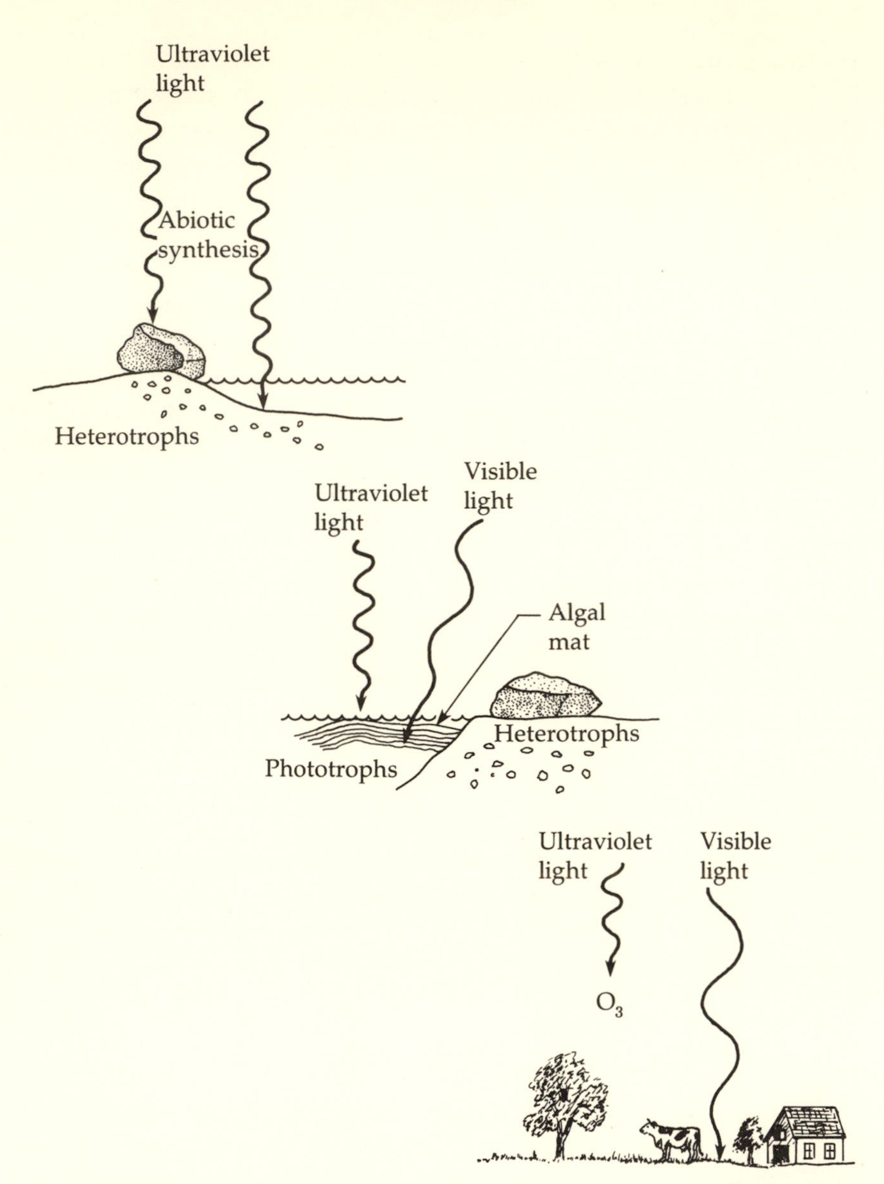

Figure 6–7. Protection against harmful ultraviolet radiation. Early heterotrophs lived beneath water, rocks, or mud. Early phototrophs lived in bacterial mats beneath a shield of dead organisms. We live beneath a screen of atmospheric ozone that developed after oxygen appeared in the atmosphere.

abiotic synthesis of food for the first heterotrophs. Phototrophic organisms, in contrast, had to live in the sunlight in order to practice photosynthesis. They had to be exposed to light at visible and infrared wavelengths, from which they extracted energy, while maintaining a screen against damaging ultraviolet radiation.

As a solution to this problem, some of the early phototrophic organisms seem to have adopted a colonial way of life. Rather than floating freely in the ocean they lived in dense matlike structures on the surface of the mud. These structures could have been a direct consequence of the postulated role of clay minerals in the origin of life. The microbes in the surface layer of the mats were killed by ultraviolet light, but the organic remains of these dead microbes absorbed the harmful radiation, providing protection for their living comrades located below the surface, while not significantly attenuating the useful visible radiation (Figure 6–7).

This notion is not as fanciful as it may appear to be. Experiments conducted at Boston University by M. Rambler and L. Margulis have shown that it works. Photosynthetic microbes in mats can thrive indefinitely, even when exposed to fluxes of ultraviolet radiation that kill the same organisms in isolation within minutes. Moreover, the fossilized remains of matlike microbial structures, called stromatolites, are abundant in ancient rocks. Indeed, among the oldest remains of undoubtedly biological origin that have yet been discovered are 3.5-billion-year-old stromatolites in rocks of the Warrawoona Group in Western Australia.

Suggested Reading

Broda, E. *The Evolution of the Bioenergetic Processes.* Oxford: Pergamon Press, 1978.

Sonea, S., and M. Panisset. *A New Bacteriology.* Boston: Jones and Bartlett Publishers, Inc., 1983.

Stanier, R. Y., M. Doudoroff, and E. A. Adelberg. *The Microbial World.* Englewood Cliffs, NJ: Prentice-Hall, Inc., 1957.

CHAPTER 7

Pollution Is as Old as Life

THE DEVELOPMENT of oxygenic photosynthesis was a substantial biochemical achievement. In order to reduce carbon dioxide in photosynthesis without using reduced reactants from the environment, organisms had to evolve some way to break the water molecule into its constituent parts, hydrogen and oxygen. The oxygen they could discard, but the hydrogen was needed for the reaction with carbon dioxide. More metabolic machinery and additional light energy was needed to split water. The additional complexity is illustrated schematically in the bottom part of Figure 6–5. Oxygenic photosynthesis uses more light than bacterial photosynthesis (approximately twice as much) to reduce a given amount of carbon dioxide.

Notwithstanding its relatively inefficient use of light, the new process offered enormous advantages to organisms. No longer was their growth limited by the supply of reduced compounds. Instead they could extract hydrogen for the reduction of carbon dioxide from the virtually unlimited supply of water. Oxygenic photosynthesis therefore made possible a great expansion in the level of biological activity.

Heterotrophs could benefit from this new development in autotrophic metabolism also, not only because it made possible an expansion in the rate of food supply, but also because the appearance of oxygen in the environment opened up an entirely new method of converting food into energy. Remember that biosynthesis consumes energy; organisms need energy in order to grow. Before the rise of atmospheric oxygen, heterotrophs obtained energy mainly by fermentation. In fermentation, one organic molecule is used to oxidize another organic molecule (see Chapter 6); the energy yield is quite limited. The appearance of oxygen made possible a more powerful energy source, aerobic respiration, in which oxygen indirectly reacts with organic compounds.

The energy released, or the work that can be done, when a given amount of sugar, for example, reacts with oxygen through a complex cycle of reactions, is more than eight times greater than the energy released by fermentation of the same amount of sugar (Figure 7–1). Compared with fermenters, respiring heterotrophs can use a much greater fraction of their food for biosynthesis, diverting less of it into energy production. A given supply of food therefore sustains more growth for respirers than for fermenters.

Oxygenic photosynthesis presented life with new problems as well as new opportunities. The waste product of photosynthesis turned out to be a dangerous pollutant because oxygen within the cell is poisonous. It reacts with certain components of the cell to produce highly reactive chemical products that indiscriminately attack any and all organic compounds. In order to survive in the presence of oxygen, cells require defense mechanisms, enzymes that catalyze the conversion of the destructive oxidants into harmless chemical compounds. Such defense mechanisms must have evolved in the first organisms to practice the kind of photosynthesis that releases oxygen as well as in the first heterotrophs to respire. Even today there are microbes that are unable to tolerate oxygen. They are obligate anaerobes, living only in restricted environments, such as the interiors of larger organisms, from which oxygen is excluded.

Unrestrained dumping of unwanted oxygen caused the most severe pollution episode in the history of the earth. The chemical environment was profoundly and permanently changed. In my opinion this change raises serious questions about the Gaia hypothesis, propounded by James Lovelock and Lynn Margulis, that the biota regulate the chemical and physical properties of their environment in such a way as to optimize the habitability of our planet. It may be possible to argue that the aerobic world is better for life than the anaerobic world that preceded it, in the sense that it can sustain a greater mass and diversity of living

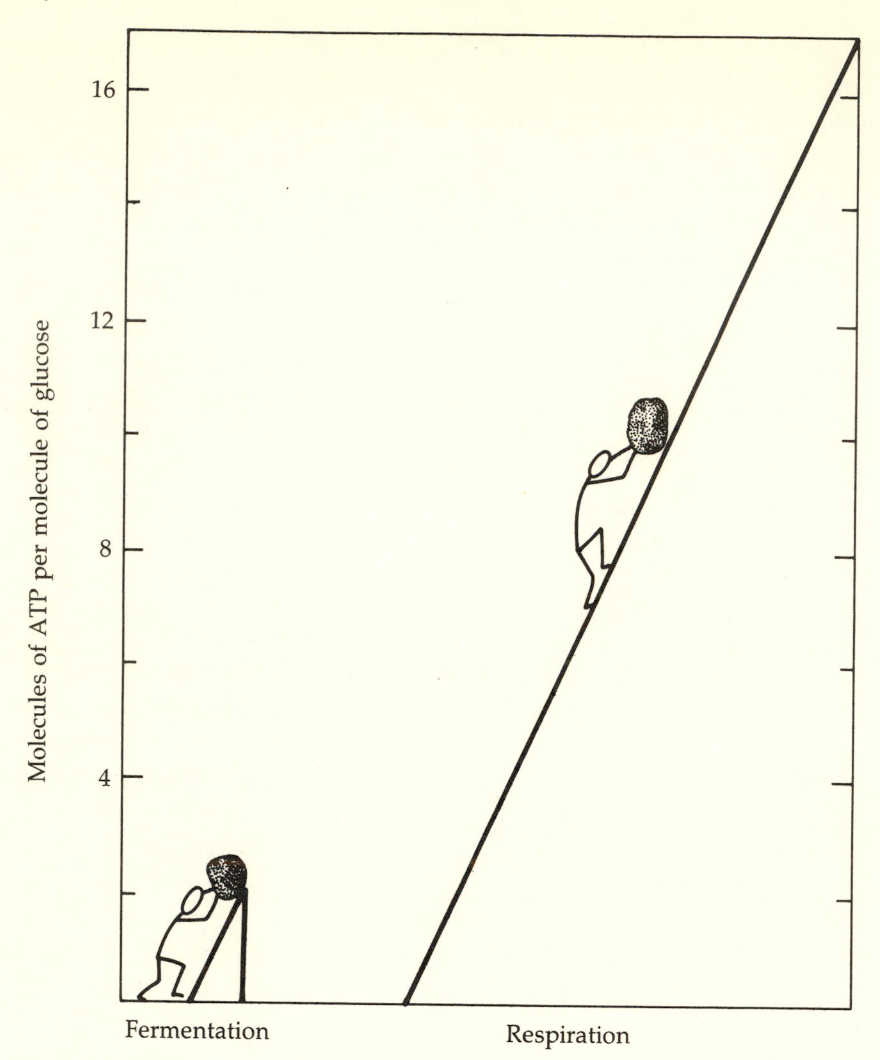

Figure 7–1. Respiration yields more energy than fermentation from a given amount of food.

creatures, but such an argument is likely to appeal only to the creatures that survived the change. The hypothesis of regulation is plainly falsified by the biologically driven transition from an anaerobic to an aerobic world. What do we know about the nature and timing of this transition and about the factors that allowed it to occur?

Oxygenic photosynthesis is practiced today by green plants

(metaphyta) and by single-celled phytoplankton (free-floating plants) that, like the metaphyta, have eukaryotic cells. But the eukaryotic cell is a later biological development than the prokaryotic cell. More pertinent to the origin of oxygenic photosynthesis are the widely distributed, simple, and apparently primitive prokaryotic microbes called blue-green algae or bacteria (blue-green refers to their pigment). Fossilized remains of blue-green bacteria (technically cyanobacteria) have been found by H. J. Hofmann of the University of Montreal in 1900-million-year-old rocks in the Belcher Islands of Canada, and some much older fossils are similar in appearance. Evidently cyanobacteria have a long history. They may, in fact, have been the organisms that initiated the "oxygen revolution."

In spite of this tentative identification, it is not known when oxygen-evolving photosynthesis arose because microfossils reveal very little about the metabolic processes of their living forms. Even when the organisms must have been photosynthetic, as in certain stromatolites, it is not yet possible to decide whether or not they released oxygen. Thus the biological record is not very informative about the timing of the oxygen revolution.

The sedimentary rock record is much better. It was interpreted originally by Preston Cloud, now at the University of California at Santa Barbara, in a way that set a date of about 2 billion years ago for the rise of oxygen. Subsequent work has added detail to the history without changing Cloud's conclusion in any basic way. The interpretation of the rock record depends on the fact that the solubility of some elements in water depends strongly on their state of oxidation. The most important example of this property is iron. In its reduced, ferrous form it is moderately soluble, but the oxidized, ferric form is virtually insoluble. Iron therefore dissolves in reducing waters (waters that are devoid of oxygen) and can be transported by them in solution. Oxidizing waters (waters containing dissolved oxygen), however, can transport iron only in the form of solid particles.

A fairly common type of rock in the Archean and Early Proterozoic was the banded iron-formation. Banded iron-formations of Proterozoic age are the major sources of iron ore today. They are chemical sediments, which means that their mineral components were precipitated from solution, and they are rich in iron. The clear implication is that they were deposited from reducing waters. Such waters, which are found today only in restricted environments, must have been widespread in the Early Precambrian. The striking feature of banded iron-formations is that they effectively disappear from the geological record about 1.7 billion years ago (Figure 7–2). Infrequent younger occurrences are qualitatively different in mineral composition and much smaller in size. A reasonable interpretation is that the time when banded iron-formations ceased to form marks the time when the ocean

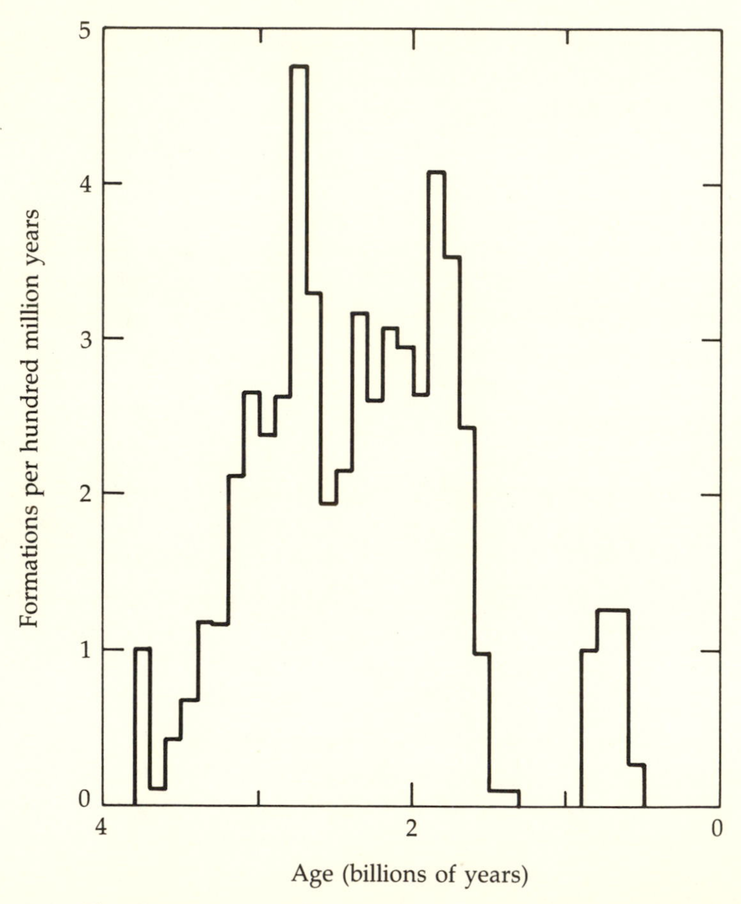

Figure 7–2. The distribution of Precambrian banded iron-formations as a function of time. (Data from J.C.G. Walker *et al.*, "Environmental Evolution of the Archean-Early Proterozoic Earth." In J.W. Schopf (Ed.), *Earth's Earliest Biosphere: Its Origin and Evolution*, pages 260–290. Princeton University Press, Princeton, NJ, 1983.)

became predominantly oxidizing. The 1.7-billion-year age of this event is uncertain because many banded iron-formations are poorly dated.

Further evidence is provided by a sparse record of ancient soils called paleosols. Modern soils generally contain more iron per unit of mass than the underlying rock because oxidizing rainwater washes soluble elements away and leaves the insoluble iron in the soil. This enrichment is absent in paleosols more than about 2.1 billion years old. Indeed, the older paleosols are deficient in iron, suggesting that they were formed by the weathering of rock in the absence of oxygen. Among the examples of soils deficient in iron that are not much older than 2.1 billion years, a few have a thin crust of oxidized iron (hematite) right at the level that was the surface at the time the soil was formed. They suggest that the atmosphere contained enough oxygen to oxidize a thin layer of iron at the soil surface, but not enough to oxidize the iron at depths within the soil.

Sandy riverbank, floodplain, and desert deposits today have a reddish pigmentation if they are derived from the erosion of iron-bearing rock. The color arises from a thin coating of hematite (Fe_2O_3) on the surface of the grains, produced by the reaction of iron minerals with atmospheric oxygen. The sedimentary rocks formed from such deposits are called red beds. Red beds are common throughout the geologic record back to about 2.4 billion years ago, but not earlier. Older rocks that appear to have formed by the same erosional processes in the same environments lack the red coloration. This evidence suggests that red beds did not form before the atmosphere contained enough oxygen to produce a thin hematite coating on the surfaces of mineral grains during the course of weathering, erosion, and transport. The ages of many of the older deposits are not well known, so the 2.4-billion-year age for this transition is uncertain, but the evidence from red beds, paleosols, and banded iron-formations can be combined to reveal a gradual increase of oxygen (Figure 7–3). At first, oxygen levels in the atmosphere were high enough to produce a hematite dust on the surface of mineral grains, but not high enough to arrest the dissolution of iron in soils. Some time later atmospheric oxygen achieved concentrations large enough to affect the composition of soils, but the hydrosphere remained overwhelmingly anoxic. The transition in the ocean was delayed for perhaps 700

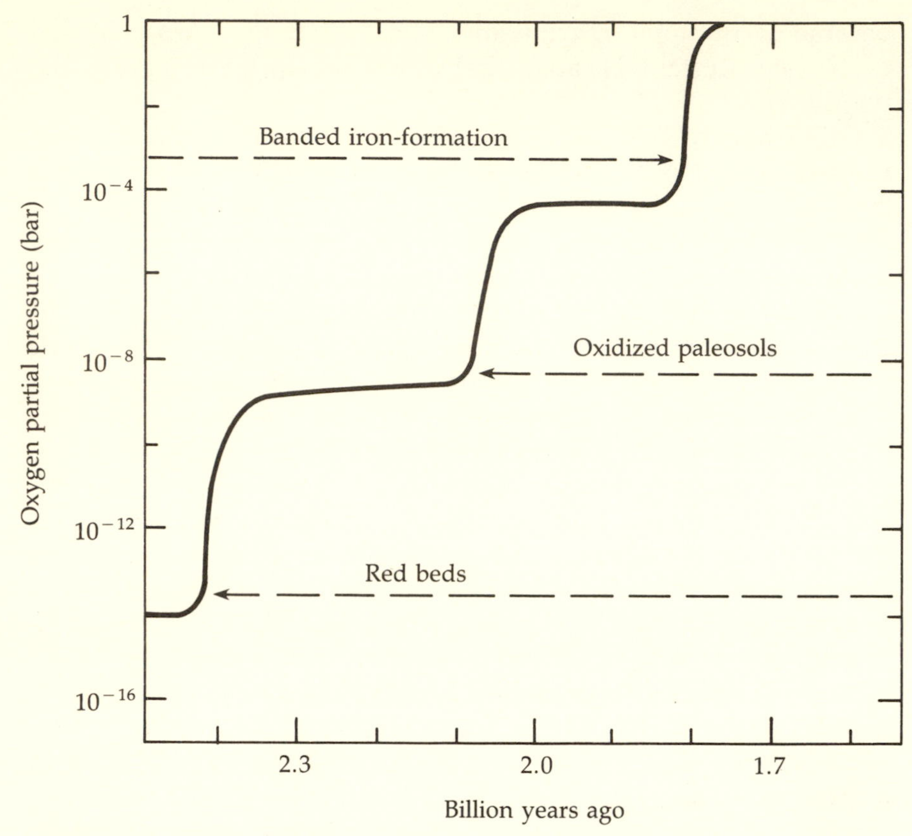

Figure 7–3. The rise of atmospheric oxygen as revealed by the geologic record.

million years after the rise of oxygen in the atmosphere first had a perceptible effect on the products of weathering and erosion.

Does this apparently extended transition make sense? It does, as consideration of the factors controlling the transition will show. Imagine a world in which the rate of oxygen production is small (Figure 7–4). Photosynthetic organisms must live in the light, so they release oxygen either directly to the atmosphere or right at the top of the ocean. Any oxygen released to the atmosphere reacts immediately with the reduced gases, principally hydrogen, released to the atmosphere by volcanoes and metamorphism. The effect of the oxygen source is to lower the concentration of hydrogen in the atmosphere, but oxygen cannot accumulate as long as the oxygen source is smaller than the

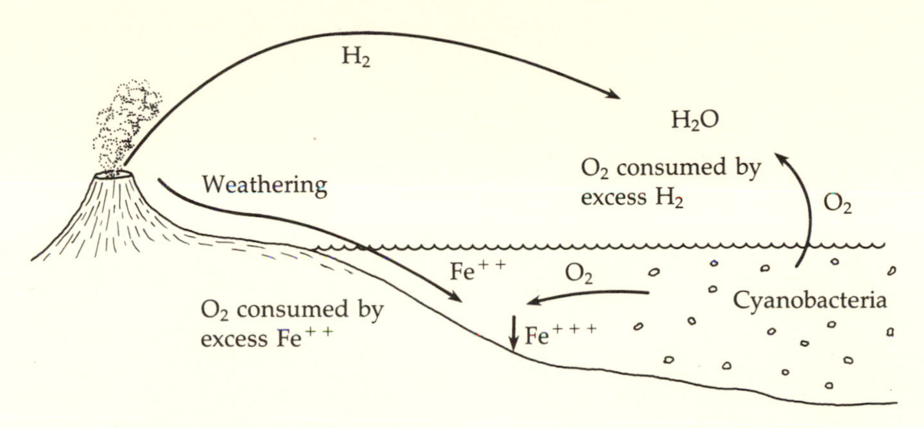

Figure 7–4. Oxygen in the anaerobic world after the origin of oxygen-producing cyanobacteria. (Adapted from J.C.G. Walker, *Origins of Life, 10*, 93–104, 1980.)

source of reduced gas. It is clear that the rise of atmospheric oxygen need not be an immediate consequence of the origin of oxygenic photosynthesis. Oxygenic photosynthesis may have originated long before oxygen first appeared in significant concentration in the atmosphere. The rise of oxygen dates from the time when the photosynthetic oxygen source first exceeded the volcanic and metamorphic source of reduced gas.

The situation in the ocean is similar. So long as the oxygen source is small, any oxygen released in the surface layers reacts immediately with reduced minerals in solution provided to the ocean by weathering reactions. The most abundant reduced species would have been ferrous iron. Oxygen cannot accumulate in the ocean as long as its rate of supply is smaller than the rate of supply of dissolved ferrous iron.

Estimates of the weathering rates and rates of volcanic emission on the early earth are very uncertain, but it seems likely that the rate of supply of dissolved reduced minerals to the ocean would have been hundreds of times larger than the rate of supply of reduced gases to the atmosphere. Suppose, then, that the photosynthetic oxygen source increased gradually with time. The first transition would have occurred when the rate of supply of oxygen to the atmosphere rose above the rate of supply of reduced gases. The transition would have been marked by a rapid decline in atmospheric hydrogen and a rapid rise in atmospheric

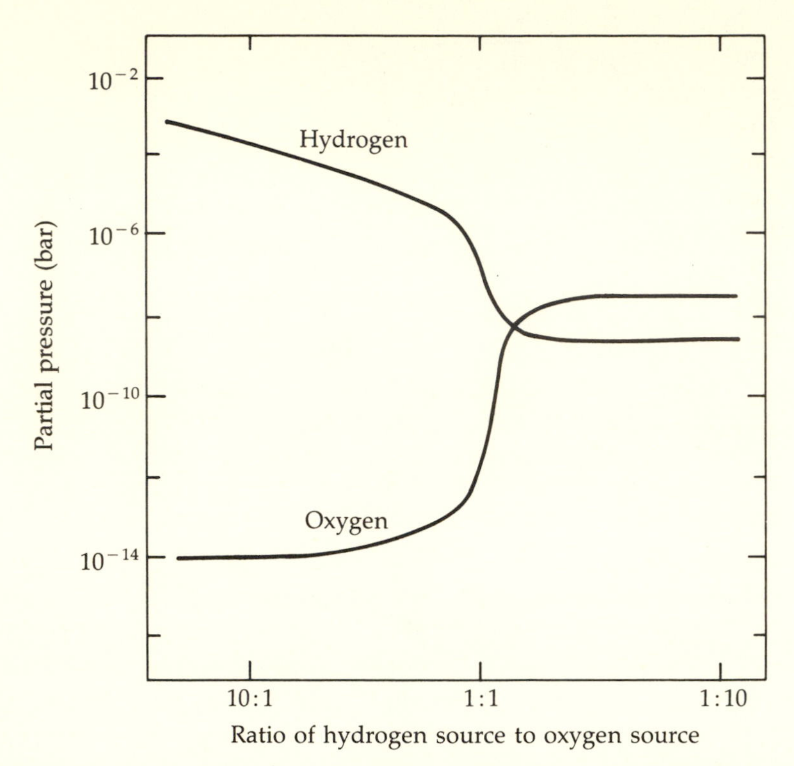

Figure 7–5. Changes in the proportions of oxygen and hydrogen in the atmosphere as the ratio of their sources passes through unity.

oxygen (Figure 7–5). The rise of oxygen would have been limited by loss to the ocean. Excess oxygen dissolved in sea water and reacted with abundant reduced minerals in solution. I have collaborated with J. F. Kasting of the National Center for Atmospheric Research on simple calculations of the atmospheric composition before and after this transition. Our results, shown in Figure 7–5, are entirely consistent with an identification of the transition as the time when red beds first appeared in the sedimentary rock record. Red beds would not have formed in the atmosphere containing excess hydrogen that we infer for the time before the transition. They could have formed in the atmosphere with excess oxygen that followed the transition.

Nevertheless, the oxygen concentration that we infer when red beds first began to form was quite low, a million times smaller than now. Increasing rates of photosynthetic oxygen production would have yielded a proportional increase in oxygen concentration, but loss of oxygen to the ocean was too rapid to permit much larger concentrations as long as the ocean remained overwhelmingly reducing. The next important transition was delayed until the rate of release of oxygen rose above the rate of supply of reduced minerals in solution. When this occurred, reduced species were eliminated from the ocean by reaction with excess oxygen, and oxygen accumulated in both atmosphere and ocean. Deposition of banded iron-formation came to an end once dissolved oxygen collected in the ocean.

Suggested Reading

Holland, H. D., and M. Schidlowski (Eds.). *Mineral Deposits and the Evolution of the Biosphere.* Berlin: Springer-Verlag, 1982.

Schopf, J. W. (Ed.). *Earth's Earliest Biosphere: Its Origin and Evolution.* Princeton, NJ: Princeton University Press, 1982.

CHAPTER 8

Pollution Can Be Helpful

WHAT LIMITED the accumulation of atmospheric oxygen once its rate of production exceeded the rate of supply of reduced compounds? The answer, of course, is aerobic respiration. To see how this works, consider the biogeochemical cycle of oxygen in the modern world, illustrated in Figure 8–1. Photosynthesis produces equal numbers of molecules of oxygen and of organic matter. The schematic reaction is

$$\text{carbon dioxide} + \text{water} \rightarrow \text{organic matter} + \text{oxygen}$$

Aerobic respiration is the reverse process, consuming equal numbers of molecules of oxygen and organic matter.

$$\text{organic matter} + \text{oxygen} \rightarrow CO_2 + H_2O$$

Today, respiration consumes cell material and oxygen just about as fast as they are produced by photosynthesis. What would happen if these two rates were not in balance? Suppose, for example, that the rate of photosynthesis exceeded the rate of respiration, as must have been the case when oxygen first appeared in the atmosphere. Then oxygen would accumulate in the atmosphere and organic matter would accumulate at the surface of the earth. More organic matter and more oxygen would lead to an increase in the rate of respiration, until this rate was just large enough to balance the rate of photosynthesis (Figure 8–2). This is how the balance is maintained today. Similar processes would have halted the original accumulation of oxygen once respiration had evolved. With

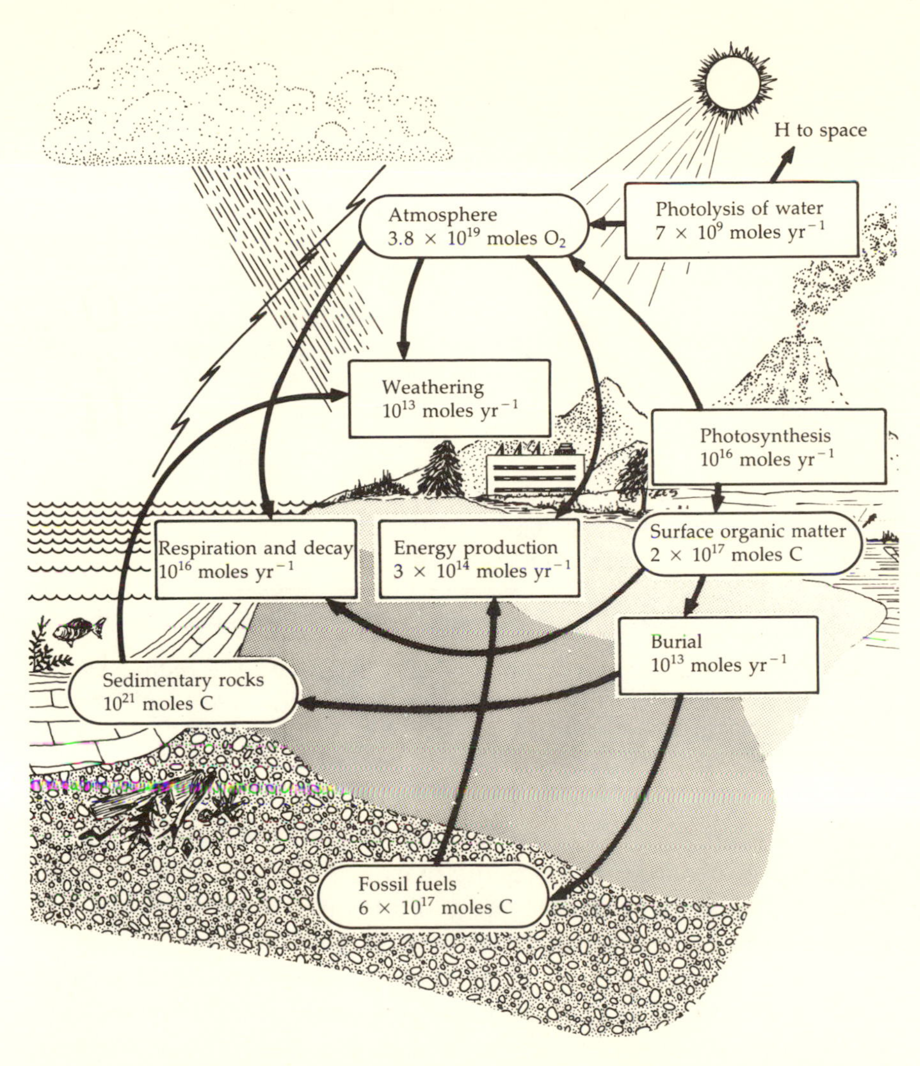

Figure 8–1. Biogeochemical cycles of oxygen in today's world. (Adapted from J. C. G. Walker, *American Journal of Science, 274,* 193, 1974.)

the onset of aerobic respiration, the atmosphere and metabolism achieved essentially their modern relationship.

Nevertheless, it is not the rapid cycle of photosynthesis followed by respiration that controls the amount of oxygen in the atmosphere. Instead, this cycle controls the amount of organic matter at the earth's surface. This conclusion follows from the

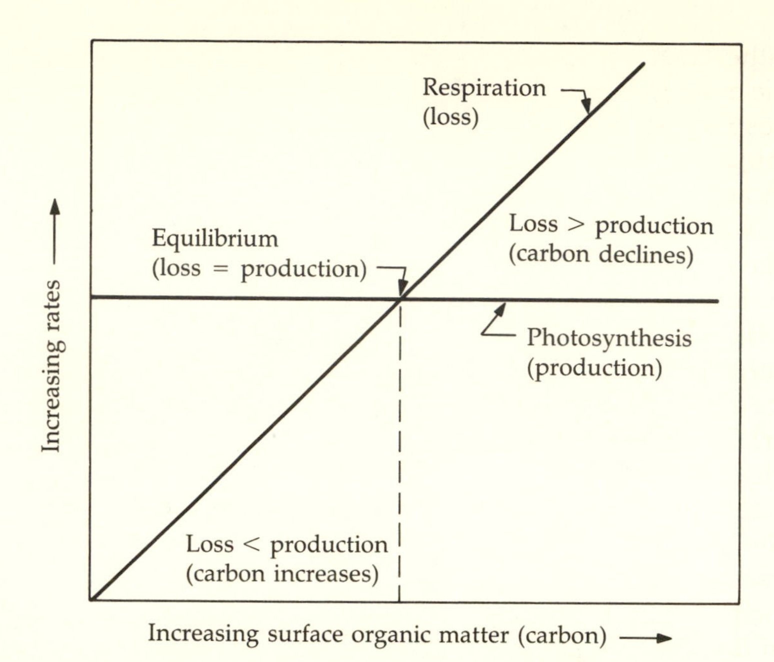

Figure 8–2. Equilibrium in the biological carbon cycle. Organic matter accumulates until the rate of respiration equals the rate of photosynthesis.

relative amounts of oxygen and organic matter. Oxygen is about 200 times as abundant as organic matter (Figure 8–1), so even if photosynthesis were to cease and all of the organic matter at the surface were to be consumed by respiration, the amount of oxygen in the atmosphere would decrease by less than 1%. The processes that control atmospheric oxygen link the atmosphere to a much larger reservoir of reduced material in the crust. Oxygen is consumed by reaction with this material when rocks are weathered and eroded. The rate of this consumption is only 0.1% of the rate of respiration, enough to exhaust all the oxygen in the atmosphere in about 5 million years.

To balance this additional consumption, a small fraction (0.1%) of the photosynthetically produced organic matter is buried in sediments rather than being recombined with oxygen in aerobic respiration. The oxygen content of the atmosphere adjusts to a level that insures that the rate of burial of organic matter in sediments just balances the rate of consumption of oxygen in weathering. Too much oxygen reduces the rate of burial of organic matter, causing oxygen to decline. Too little reduces the rate of respiration, so more organic matter is buried and oxygen accumulates. (Remember, organic matter that is not buried reacts with and consumes oxygen in aerobic respiration.)

So much for the processes that controlled the level of oxygen in the environment before, during, and after the oxygen revolution. Key questions still unanswered include: Was the transition inevitable or could the world have remained anaerobic forever? What determined the timing of the transition? Why did it take place about 2 billion years ago rather than much earlier or much later? These questions would be answered if we knew why the source of oxygen, photosynthesis, increased relative to the sources of reduced gases in the atmosphere and reduced minerals in the ocean. Note that it is not clear whether the oxygen source increased, or the source of reduced material decreased, or both. If we suppose, for the moment, that the oxygen source increased, then it is appropriate to wonder what factors could have limited the rate of oxygen production at first and then gradually relaxed to allow the increase.

One factor, suggested by Andrew H. Knoll of Harvard University, was the limited area of suitable habitats for the first photosynthetic organisms. The most biologically productive regions of the world ocean today are the continental shelves, where nutrients are relatively abundant and there are extensive areas where the water is not too deep for photosynthetic organisms to live on the bottom. In the Archean, prior to 2.5 billion years ago, large continents had only just begun to form and continental shelf environments were restricted. Biological productivity, including oxygen production, may have been correspondingly low. The increase in continental area that marked the beginning of the Proterozoic (discussed further in Chapter 10) may have permitted an increase in the rate of oxygen production by opening up more area for colonization by photosynthetic organisms.

A further suggestion that appeals to me is that the rate of oxygen production was limited, on the anaerobic earth, by competition between bacterial photosynthesis and oxygenic photosynthesis. Oxygenic photosynthesis uses more light energy than bacterial photosynthesis to make a given amount of cell material because of the need to extract hydrogen for the cell material from water. Photosynthetic bacteria have a competitive advantage if light is limiting and there is an adequate supply of the reduced compounds they require. Indeed, there are a number of species of cyanobacteria that practice bacterial photosynthesis if reduced reactants are available and switch to oxygenic photosynthesis only when the supply of reduced reactants is exhausted.

The resource for which these two forms of photosynthesis were competing may have been dissolved phosphorus. Phosphorus is an essential nutrient element without which growth is impossible. Phosphorus is supplied to the ocean by the erosion of rocks in amounts so small that shortage of phosphorus limits biological productivity over much of the modern ocean. Only in local areas where phosphorus is abundant are there massive growths of microorganisms and the larger creatures that feed upon them. These are the areas of the world's great fisheries.

We can imagine that phosphorus was in short supply in the Archean ocean also, and that oxygenic photosynthesis had to compete with bacterial photosynthesis for what there was (Figure 8–3). Because of its lower efficiency, oxygenic photosynthesis could use only phosphorus left over when bacterial photosynthesis had exhausted the supply of reduced compounds. If we further imagine that the rate of supply of reduced material, derived from volcanic hydrogen, declined with time as the interior of the earth cooled down, then decreasing competition from photosynthetic bacteria would have permitted a gradual increase in the rate of production of oxygen (Figure 8–4).

There are two suggestions, therefore, for the increasing rate of oxygenic photosynthesis. An increased area of stable shallow water environments on continental shelves provided more space for bottom-dwelling photosynthesizers, and reduced release of volcanic hydrogen caused less competition from bacterial photosynthesis. Neither of these ideas has received any quantitative analysis yet; our understanding of the underlying mechanisms of the oxygen revolution is rudimentary. For whatever reason, the rate of oxygen production rose above the rate of release of reduced compounds to the atmosphere and ocean. The biosphere was swept free of abundant reduced material by reaction with excess oxygen, and oxygen accumulated to a level sufficient to support aerobic respiration.

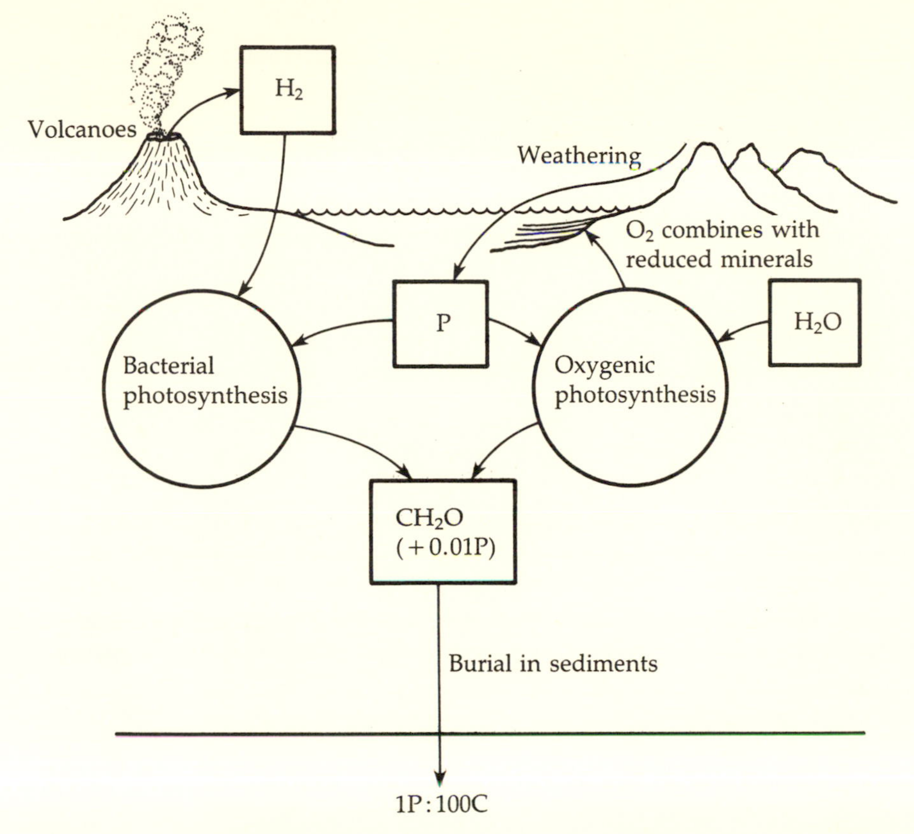

Figure 8–3. Competition between bacterial photosynthesizers and oxygenic photosynthesizers for phosphorus in the Archean ocean. (Adapted from J. C. G. Walker, *Origins of Life, 10,* 93–104, 1980.)

The onset of aerobic respiration brought to an end the era of coevolution of atmospheric composition and microbial metabolism. Since that time, perhaps 1.7 billion years ago, life and its environment have been characterized by metabolic and biogeochemical stability. Life has responded to changes in the biosphere, most of them geological in nature, but it has stopped (until very recently) causing profound changes itself. In the rest of this exploration of earth history, therefore, I shall refer to life more as an indicator of environmental change than as an initiator.

The impact of pollution by oxygen was not all bad. The great

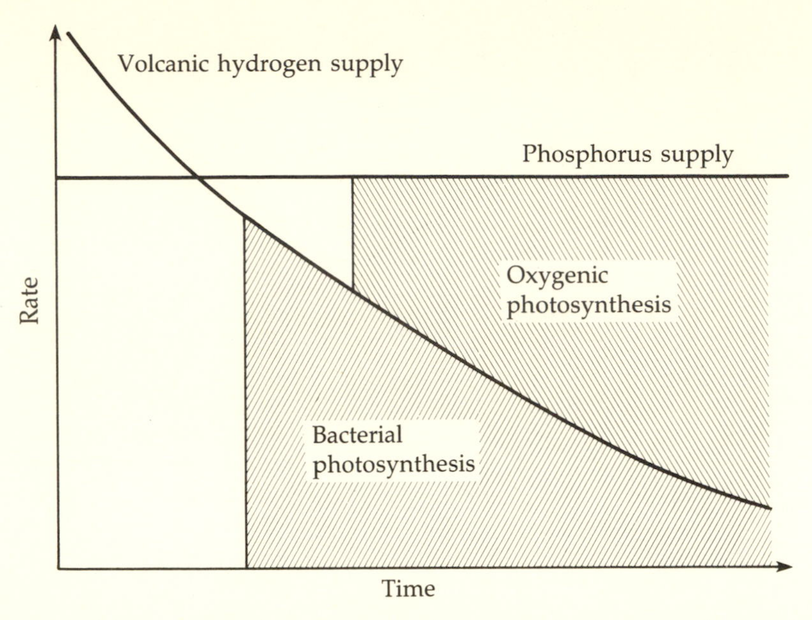

Figure 8–4. The production of oxygen by oxygenic photosynthesis might increase as the rate of supply of hydrogen decreases, because of reduced competition for phosphorus by bacterial photosynthesis.

efficiency of aerobic respiration as a source of energy permitted important biological developments. Two key innovations already mentioned must be described in greater detail. The first was the development of a new kind of cell, called the eukaryotic cell, with its genetic material contained in a cell nucleus surrounded by a membrane, and several different internal structures, called organelles, performing different special functions (see Figure 1–3). All of the microbes I have discussed so far consisted of prokaryotic cells, with much simpler internal organizations. The eukaryotic cell offered life a number of advantages, the principal one being the possibility of a sexual mode of reproduction in which genetic material from two individuals mixes to produce a new individual having only half the genes from each parent (see Figure 6–2). The increased variability thereby introduced into a sexually reproducing population accelerates the processes of evolution by producing a greater range of variants for natural selection to work on. Prokaryotic organisms reproduce asexually, by simple division of the cell, a process that, apart from the occasional mutation, produces offspring genetically identical to the parent. Thus, evolution among prokaryotes must rely entirely on mutations and, in some species, infrequent haphazard exchange of genetic material between individuals.

The origin of the eukaryotic cell is still subject to debate, but evidence has been accumulating in recent years in support of a theory invoking symbiosis, the mutually beneficial interaction of two or more different species of organisms. Cows, for example, depend on the bacteria in their rumens to break down the cellulose of grass into chemicals that the cow can digest. The rumen bacteria, in turn, benefit from a sheltered environment and a reliable supply of cellulose. According to the symbiotic theory, the internal organelles of the eukaryotic cell arose from symbioses between different kinds of bacteria. Cyanobacteria, for example, may have developed symbiotic associations with heterotrophic bacteria, in which each partner furnished nutrients to the other. In time the photosynthetic partner may have moved inside the cell of the heterotrophic partner. For example, the photosynthetic apparatus of modern eukaryotic cells is localized in intracellular organelles called chloroplasts. Chloroplasts may be relics of formerly independent cyanobacteria.

All of the organisms we see around us in the world today are composed of eukaryotic cells. Only bacteria, including cyanobacteria, are prokaryotes. They have survived competition with eukaryotes largely because of the diversity of their metabolic capabilities, and because of niches, including the insides of eukaryotes, in which they can grow. For example, essentially all eukaryotes are aerobic and cannot survive without oxygen, but prokaryotes span the range of responses to oxygen. Some of them cannot tolerate it, some cannot live without it, and some are indifferent to it. Thus prokaryotes can live in habitats denied to eukaryotes. Evidently the eukaryotic cell evolved after the world had become aerobic, while prokaryotes adapted to a world of changing oxygen concentration.

Eukaryotic cells are most easily identified by their internal structure, but this structure is seldom preserved by the process of fossilization. Indeed, experiments have shown that features resembling nuclei and other internal organelles can be generated during the fossilization of prokaryotic cells. The identification of the oldest eukaryotic fossil is therefore controversial. Fortunately, some eukaryotic cells are significantly larger than prokaryotic cells, a circumstance that permits their tentative identification in the fossil record. On the basis of statistical analysis of cell sizes in microfossil populations it appears that eukaryotes were in existence as much as 1.4 billion years ago. By that time the world must have been aerobic.

Some time after the appearance of the eukaryotic cell came organisms composed of large numbers of individual cells performing different specialized functions. For the first time organisms could develop shells, circulatory systems, and teeth. The oldest fossilized remains of multicelled animals (called metazoa) have been found in the 700-million-year-old Ediacara formation in southern Australia. The Ediacara fauna consisted of soft-bodied creatures resembling jellyfish and worms. Shortly thereafter, shell-forming metazoa became abundant, and fossils have preserved a rich record of the evolution of life ever since. It appears that neither eukaryotes nor metazoa could have originated without the abundant metabolic energy of aerobic respiration. The oxygen revolution was ultimately helpful to life.

The paucity of the fossil record throughout the first 4 billion years of earth history (the Precambrian era) is the reason why descriptions of the early earth have been so speculative. For a start, nearly all Precambrian organisms were soft-bodied and therefore only rarely were preserved as fossils. Second, all organisms known prior to Ediacaran time were single-celled microbes, and it is well-nigh impossible to tell from the fossilized remains of a microbe how that microbe behaved when it was alive. Microbes with very similar appearances frequently have very different metabolisms. Among microbes, therefore, there is little relationship between form and function; fossilization preserves only form, and even that is frequently not preserved very well. On top of all this, well-preserved Precambrian rocks are simply not as abundant as younger rocks. To reconstruct the Precambrian history of the biosphere scientists have therefore had to rely heavily on imagination guided by the record found in stromatolites, sedimentary rocks, and occasional well-preserved occurrences of microbial fossils.

Suggested Reading

Fenchel, T., and T. H. Blackburn. *Bacteria and Mineral Cycling.* London: Academic Press, 1979.

Gottschalk, G. *Bacterial Metabolism.* New York: Springer-Verlag, 1979.

Margulis, L. *Origin of Eukaryotic Cells.* New Haven: Yale University Press, 1970.

CHAPTER 9 *The Not-So-Solid Earth*

TWO SOURCES of change have been active in earth history since life first appeared. One is biological evolution. Early changes in the forms of life and their impact on the environment have already been described, and more will be said about biological evolution in later chapters. An equally important evolutionary force is tectonic activity—the restless motion of different portions of the earth's surface. Tectonic activity has led to the deformation and melting of rocks, the eruption of volcanoes, mountain building, and relative motion of the continents. These geological processes are driven by gravity and by the escape of heat from the earth's interior, which causes a slow convective overturning of the mantle (the thick layer of silicate rock that lies between the core and the crust). This chapter describes the patterns of tectonic change; later chapters explore the impact of tectonic processes on earth history.

Crustal evolution is an important aspect of earth history in its own right. It is also important because of its impact on the course of biological evolution. This impact is most clear in the case of the multicelled animals and plants whose history is described in later chapters. In addition, it is more than likely that the emergence of continental land masses influenced biological evolution during the microbial era that has already been described. As suggested in Chapter 8, biological productivity and the rate of release of photosynthetic oxygen may at one time have been restricted by the absence of stable shallow water environments before the continents developed.

The easiest way to deal with the history of the crust is to begin with the present. The present situation is the one that is best known and can be described most precisely. Then a description of earlier conditions, which must become increasingly speculative and vague as age increases, can be based on comparisons with the present.

The surface layer of the solid earth is called the crust (Figure 9–1). There are two kinds of crust. That under the oceans is about 7 km thick on the average and is composed of relatively dense rock, described as basaltic, that is rich in magnesium and iron. Continental crust is much thicker, about 40 km on the average, and consists largely of less dense granitic rock, rich in silica and aluminum. Underlying the crust is the upper mantle composed of rocks that are even denser and richer in magnesium and iron than the basaltic rocks of the oceanic crust.

Continental and oceanic crust may be thought of as floating on top of the denser rocks of the upper mantle. The surfaces of the continents rise above the level of the sea floor because continental crust is thicker and composed of less dense material than oceanic crust. In the same way, a thick slab of cork floats higher than a thin plank of wood. If the continental crust were planed down to a thickness comparable to that of oceanic crust the continents would not rise sufficiently above the sea floor to keep sea water from flooding the land.

Temperatures increase with depth within the earth, so that

BASALTIC AND GRANITIC ROCKS

The distinction between basaltic and granitic rocks is one of chemical composition and density. The rocks of the continental crust have an over-all chemical composition resembling that of granite. They are rich in silica and potassium and relatively deficient in iron, magnesium, and calcium. Their density is relatively low. Oceanic crust is composed largely of basalt, richer in iron, magnesium, and calcium, poorer in silica, and denser. The rocks of the upper mantle are denser still. They contain less silica and calcium than basalt, more iron, much more magnesium, and much less aluminum.

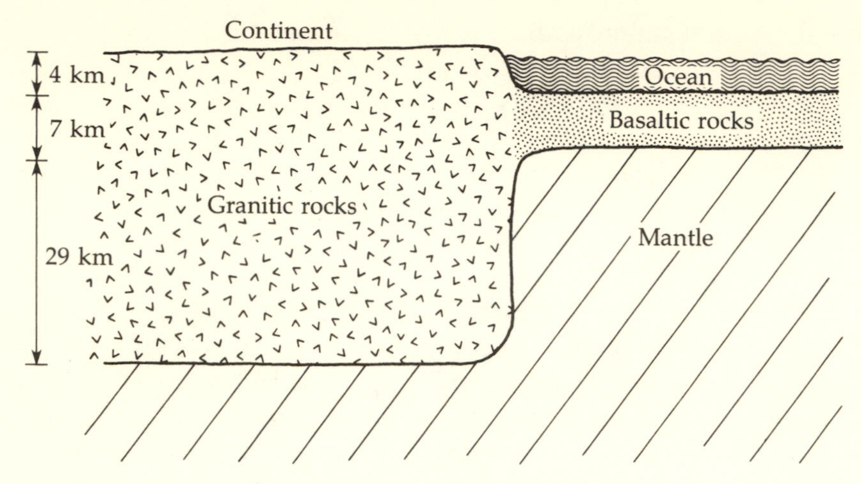

Figure 9–1. Earth's crust is thicker under the continents than under the oceans. Continental rocks, rich in silica and aluminum, are less dense than the rocks of the oceanic crust, which are rich in magnesium and iron.

rocks at greater depths are plastic. At a level within the mantle called the asthenosphere rocks lose their strength and acquire the ability to flow. Overlying the asthenosphere is a cold, rigid layer called the lithosphere. With a thickness varying from a few tens to 100 km under the ocean and perhaps 200 km under the continents, the lithosphere includes the crust and the top of the mantle.

According to the modern theory of plate tectonics, developed since 1965, the lithosphere is broken up into about a dozen major plates and many smaller fragments that together cover the surface of the globe (Figure 9–2). These plates move with respect to one another, sliding over the lubricating layer of the asthenosphere. The continents are carried along by this motion, embedded in the surfaces of much larger plates; thus the theory of plate tectonics incorporates the much older theory of continental drift. The most obvious manifestations of tectonic activity, earthquakes and volcanoes, occur mainly along the boundaries between adjacent plates.

The boundary between plates that are moving away from one another can be described as an axis of separation or spreading and is usually marked by a midocean ridge. The gap in the lithosphere left by the separating plates is filled by the upwelling of

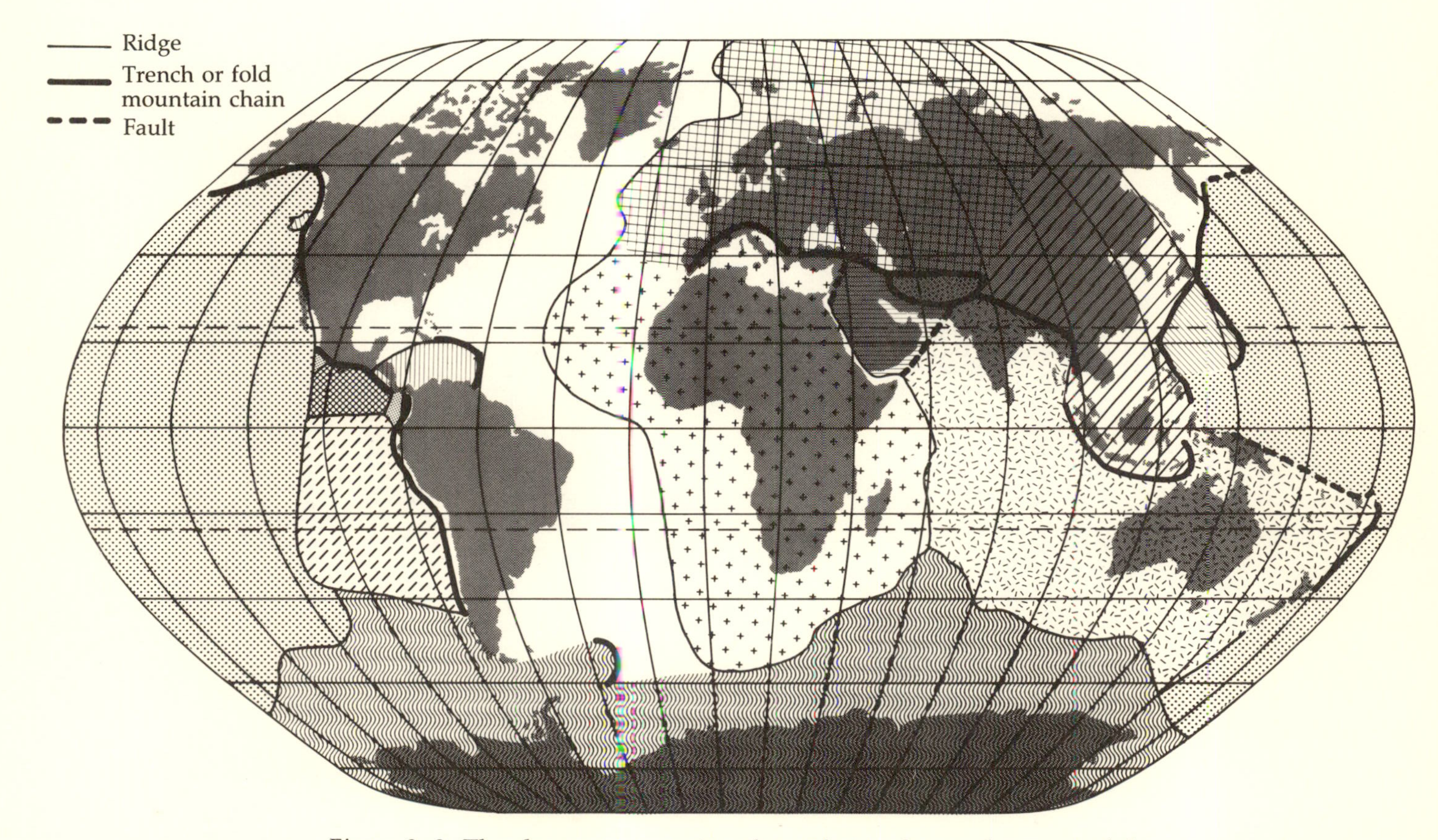

Figure 9–2. The dozen or so major plates that make up the crust of the earth. (Adapted from W. J. Morgan, *Journal of Geophysical Research, 73,* 1959, 1968.)

molten rocks derived from the mantle. Upon solidification, these rocks, which are basaltic in composition, become new oceanic crust. From time to time a new axis of spreading appears in the interior of an existing plate, so the plate is broken into two portions moving apart. If the break happens to cross a continent, the two parts of the continent gradually become separated by a new ocean. The Atlantic Ocean was formed in just this way. Separation began about 150 million years ago along what is now the Mid-Atlantic Ridge and has continued to this day. As a result, the width of the Atlantic Ocean increases by a few centimeters every year. The Red Sea and the East African Rift Valley were formed by an axis of spreading that has only recently become active. In time a new ocean may separate Africa from Arabia.

The separation of plates along midocean ridges results in the formation of about 5 square kilometers of new crust every year. Because the circumference of the earth is not getting larger, its surface area must remain constant. There must therefore be places at which crust is consumed at a rate that equals crust formation. These places are found at the boundaries between plates that are moving toward one another, called subduction zones. In a subduction zone the lithosphere of one plate bends sharply downwards and slides, at an angle of about 45°, into the asthenosphere beneath the other plate (Figure 9–3). As it descends, the material of the descending slab is gradually heated to the temperature of the surrounding asthenosphere. It becomes soft, begins to flow, and loses its identity. In this way, crustal material is returned to the mantle to replace the mantle material used to make new crust at midocean ridges. Sea floor sediments rich in water and other volatiles are carried downwards on the descending slab. The presence of water gives this material a low melting point, so much of it returns to the surface by way of volcanoes. Subduction zones are therefore frequently marked by chains of volcanic islands, called island arcs. The Aleutians are an example. The oceans attain their greatest depths in the trenches that lie along the lines where one lithospheric plate turns downwards under another.

The trailing edge of a continent (the edge that is moving away from an axis of sea-floor spreading) can accumulate a great thickness of sediments, called a geosyncline, derived by erosion of the land surface (Figure 9–4). The east coast of North America

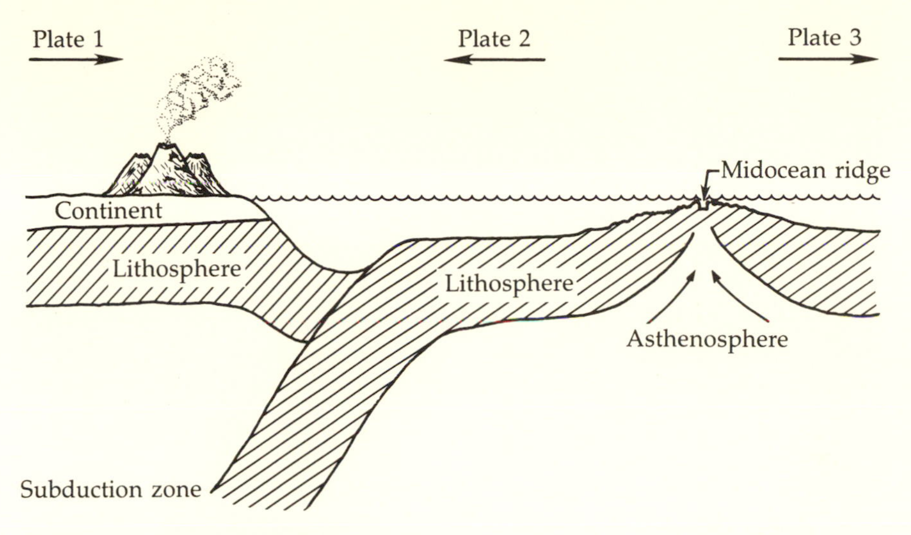

Figure 9–3. Plate movement.

provides an example. Geosynclines are the precursors of mountain ranges. A changing pattern of plate movement can convert the trailing edge of a continent into the leading edge. In time the continent and its associated geosyncline will encounter a subduction zone. Continental rocks and the sediments derived from them are too light to be dragged down into the mantle in large quantities. Instead, the sediments of the geosyncline are crumpled up into a mountain range, metamorphosed, and even melted in part to yield volcanoes.

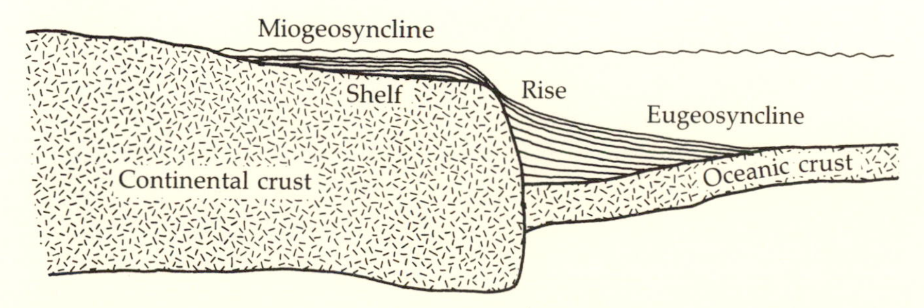

Figure 9–4. Sedimentary geosynclines accumulate at the junction between continental crust and oceanic crust.

The details of mountain building, of course, depend on the particular circumstances of the collision between continent and subduction zone. Upon occasion, for example, two continents can be brought together by subduction of the oceanic crust that once separated them. The Himalayan Mountains were formed by such a continent–continent collision that began about 45 million years ago.

Erosion is a fairly rapid process, in geological terms, that has continued throughout earth history. It tends to lower the elevation of the land surface to sea level. The building of mountains from geosynclinal deposits of sedimentary rocks performs the very important function of getting the products of erosion off the bottom of the sea and back onto the land. The effect of erosion is to spread continental (granitic) material uniformly over the surface of the globe. The effect of sea-floor spreading and mountain building is to sweep this material back into the continents. The difference in thickness between continental crust and oceanic crust is therefore a consequence of tectonics. Similar distinctions between different kinds of crust with different thicknesses are not apparent on the other planets (see Figure 9–5). In terms of this theory, we expect mountains to be built at the edges of continents; this expectation is well-supported by the geologic record. Except in cases where two continents have joined to form one, interiors of continents tend to consist of stable platforms that have been protected from major deformation for long periods of time. The platforms have not been far below the surface of the sea since they first formed. They have accumulated only a few thousand feet of sedimentary rocks compared with the tens of thousands of feet that have accumulated in geosynclines. They have not been deeply eroded either, although numerous breaks in the sedimentary record indicate that they have frequently been above sea level. Evidently the platforms have remained at elevations close to sea level throughout much of their history, while large vertical movements of the crust have occurred mainly at the continental margins.

The theory of plate tectonics constitutes a major revolution in scientific understanding of earth history. It has brought under a single conceptual umbrella a mass of previously unrelated observations in geology, geophysics, and paleontology. Its origins lie in the theory of continental drift, which was developed and

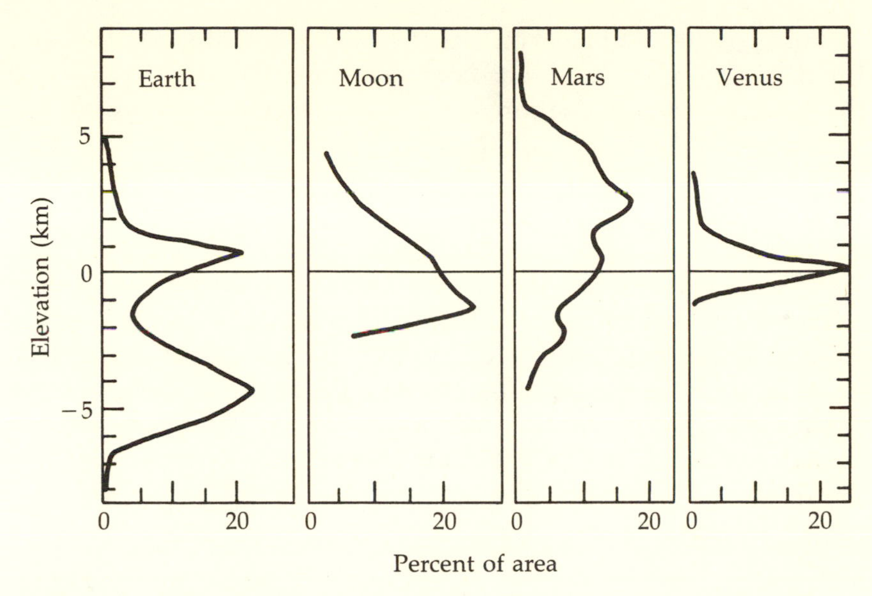

Figure 9–5. The distribution of elevations on the earth (plotted as a percentage of the total globe at a given elevation) shows two maxima, in marked contrast to the other planets. This distribution reflects the distinction between continental and oceanic crust on Earth. (After H. Masursky et al., *Journal of Geophysical Research, 85,* 8232, 1980.)

defended most vigorously during the first decades of the twentieth century by the German astronomer and meteorologist, Alfred L. Wegener. Wegener's ideas received little support after his death in 1930 (while he was leading an expedition of exploration in Greenland), mainly because earth scientists found it hard to understand how the continents could move through the rigid floor of the ocean basins.

Opinions began to change in 1956 as a result of measurements made by a group of geophysicists in England who were studying paleomagnetism. When an igneous rock cools and solidifies in the magnetic field of the earth, the iron-bearing minerals it contains acquire a small magnetization, like the magnetization of a compass needle, aligned parallel to the magnetic field of the earth. By measuring the direction of this remanent magnetization, as it is called, it is possible to determine the direction of the earth's magnetic field, relative to the rock being sampled, at the time of formation of the rock. When such data are compiled for rocks of different ages on a particular continent it appears that the direction of the field has changed with time. Comparison of the remanent magnetization of rocks of the same age on different continents reveals different apparent directions of the field. These observations suggest that the continents have been moving, both with respect to one another and with respect to the earth's magnetic field.

While the British workers were clarifying the paleomagnetic evidence of continental drift, Maurice Ewing and Bruce Heezen of Columbia University reported that the midocean ridges form a continuous system extending through all of the oceans, making them the largest mountain range in the world. The significance of the ridges as centers of sea-floor spreading was recognized by H. H. Hess of Princeton University in about 1960, and the reality

MAGNETIC ANOMALIES ON THE SEA FLOOR

The magnetic field of the earth reverses direction from time to time. As new sea floor is formed by the upwelling and solidification of hot basalt at a midocean ridge, it acquires a remanent magnetization in the direction of the field (Figure 9–6). This remanent magnetization is preserved as the newly formed crust moves outwards from the ridge. The result is a pattern of stripes, symmetrical about the ridge, of remanent magnetization alternately in one direction and the other. This pattern can be mapped with magnetometers towed behind ships. The interpretation of such data placed the theory of plate tectonics on a firm foundation.

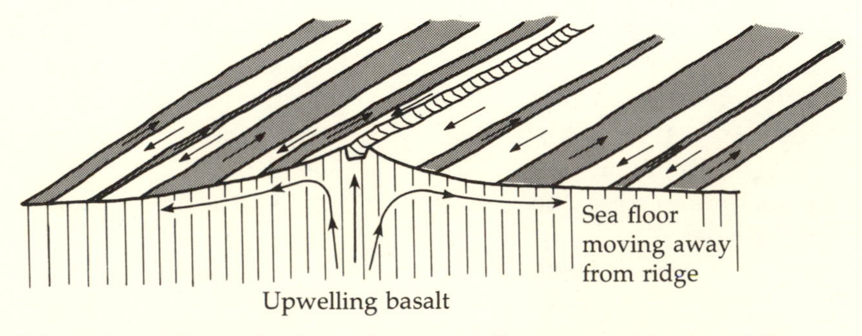

Figure 9–6. Magnetization of new sea floor at a midocean ridge.

of sea-floor spreading was established within the next 5 years as a result of an interpretation, suggested by F. J. Vine and D. H. Matthews of Cambridge University, of a perplexing pattern of anomalies in the remanent magnetization of the rocks of the sea floor.

The concept of sea-floor spreading eliminated the problem, faced by the older hypothesis of continental drift, of how the continents move through the rigid sea floor. We now know that continents and sea floor move together, from midocean ridge to subduction zone. Once clear evidence of sea-floor spreading had been obtained, the full theory of plate tectonics developed rapidly as a result of the work of many different geologists and geophysicists. Today it is almost universally accepted.

Suggested Reading

Dott, R. H., and R. L. Batten. *Evolution of the Earth.* New York: McGraw-Hill Book Co., 1976.

Scientific American. *Continents Adrift and Continents Aground.* San Francisco: W. H. Freeman and Co., 1976.

Windley, B. F. *The Evolving Continents.* New York: John Wiley and Sons, 1977.

CHAPTER 10

What Came Before Plates?

ALTHOUGH nearly all earth scientists now believe that the theory of plate tectonics correctly describes geologic processes on the modern earth, there is little agreement over whether plate tectonics in its modern form has always been the dominant process shaping the face of the globe. The problem is that some investigators find evidence for tectonic processes in the distant past (the Precambrian era) that were markedly different from those of the last six hundred million years, while other investigators feel that any differences are more of degree than of kind. The debate touches one of the most deeply respected precepts of geology, the Principle of Uniformitarianism, propounded by the 18th-century Scottish physician and farmer, James Hutton. This theory became solidly entrenched in geological thinking during the course of the nineteenth century, largely as a result of the efforts of Charles Lyell. It holds that geological processes that operated in the past were no different from the processes that are still operating, so past events can be interpreted in terms of processes that we can observe, without recourse to special circumstances or spectacular catastrophes (Noah's flood, for example).

The Principle of Uniformitarianism has been a cornerstone of geological theory for more than a century, but during this time geological research has been concerned mainly with the Phanerozoic era. It is now becoming apparent that the Principle may not apply strictly to the earlier, much longer Precambrian era. Indeed, in previous chapters I have discussed the primitive anaerobic atmosphere without a thought for uniformitarianism. For example, weathering processes in an oxygen-free atmosphere can no longer be observed on Earth. This does not mean that they never occurred, whatever the Principle of Uniformitarianism might imply. Like our aerobic atmosphere, plate tectonics may not have been present during all of earth history.

Tectonic activity results from the transport of heat from the hot interior of a planet to the cold surface. The nature of tectonic activity depends on the magnitude of the flux of heat, internal temperatures, the way in which the mechanical properties of planetary material vary with temperature and pressure, and the gravitational field of the planet. None of the other planets shows recognizable evidence of plate tectonics. With the possible exception of Venus, about which there are not enough data, the other planets do not exhibit the clear distinction between thick continental crust of relatively light rocks and thin oceanic crust of dense rocks. Any light crustal material on these planets is spread around more or less uniformly; it has not been concentrated into continents by horizontal plate motions.

Because Moon, Mercury, and Mars are much smaller than Earth, they are believed to have colder interiors. A small planet is heated less by accretion and the decay of radioactive minerals, and it cools more quickly because it has a large surface area relative to its volume. Cool interiors produce thick lithospheres, too strong to be broken into plates and moved around the surface of the globe. Heat escapes from the interiors of these planets mainly by conduction through the lithosphere. In the past, when heat flows were larger, there were episodes of volcanism during which hot molten magmas escaped from the interiors to the surface (the lunar maria and the mountains on Mars are examples). For Venus, however, a different tectonic style cannot be attributed to smaller size because Venus and Earth are the same size. Debate continues about tectonic patterns on Venus.

The thick stable lithospheres of the smaller planets have preserved a record of events in the distant past that has been obliterated on Earth by erosion and continuing tectonic activity (Figure 10–1). The ages of lunar rocks extend all the way back to the time of formation of the solar system. Moon, Mars, and Mercury have heavily cratered surfaces that reflect bombardment by meteoritic material during the final stages of accretion. Studies of lunar rocks

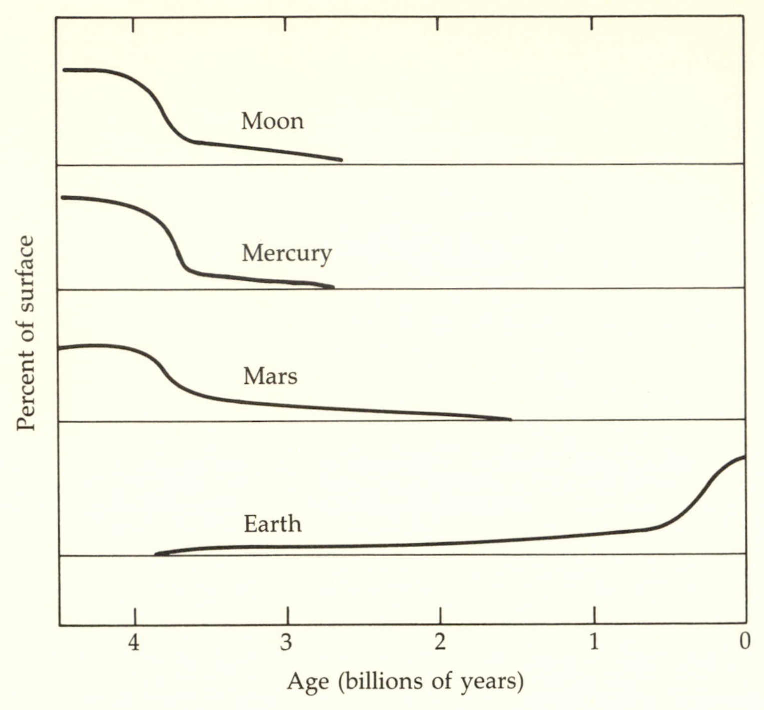

Figure 10–1. Ages of planetary surfaces.

indicate that the rate of bombardment was large until about 3.9 billion years ago. Earth must have been subject to this same bombardment, so the earliest tectonic activity on Earth would have been dominated by accretion and the formation of impact craters rather than by internal processes.

The lunar evidence indicates that meteorite bombardment became relatively infrequent after about 3.9 billion years ago, which is only a short time before the formation of the oldest terrestrial rocks yet discovered (at Isua in western Greenland). Calculations indicate that the interior of the earth was hotter then than it is now, and that the rate of generation of heat by the decay of radioactive minerals was larger by a factor of three or four. Heat flow from the interior to the surface must have been larger by a factor at least as great. If this large amount of heat was carried to the surface as heat is today—by a plate tectonic mechanism (hot material emerges along axes of spreading and cold material descends into the interior in subduction zones)—there must

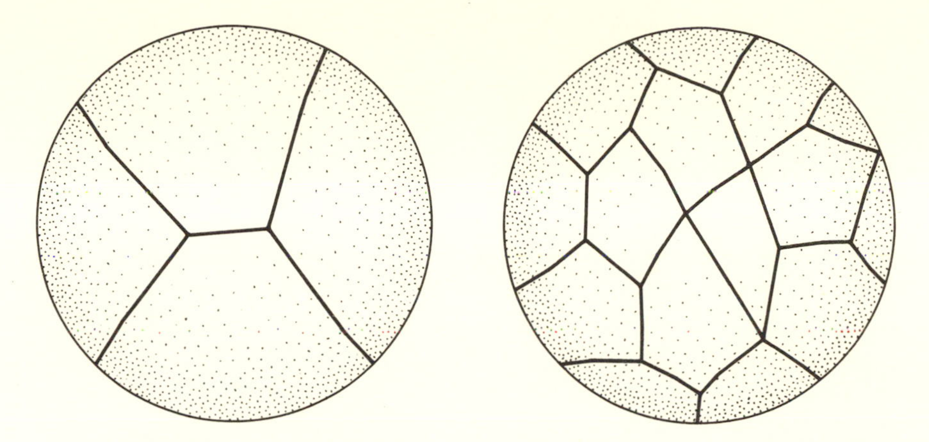

Figure 10–2. Globes divided into a few large plates by a relatively short total length of spreading axes and into many small plates by a greater total length of spreading axes.

have been many more axes of spreading or the motions must have been faster, or both. Paleomagnetic data suggest that rates of continental drift were not unusually large in the Precambrian, so it appears that there must have been more axes of spreading. A four-fold increase in heat flow requires a four-fold increase in the combined length of axes of spreading, if velocities remain unchanged. This implies a 16-fold increase in the number of plates into which the surface of a globe of constant area is divided by spreading axes and associated subduction zones (Figure 10–2). Thus, if something like plate tectonics was operating on the early earth, we could expect a couple of hundred small plates in place of the dozen or so major plates of the modern world.

AREA AND LENGTH

The number of plates of approximately equal area into which the fixed area of the globe can be divided is proportional to the square of the length of the dividing line. To see this, take a sheet of paper and fold it once. The fold is the dividing line (spreading axis), and the two halves represent plates. Fold it again, the other way, and count the layers. There are four layers (plates) and two dividing lines. Fold it another time. There are three folds and eight layers. A fourth fold gives 16 layers.

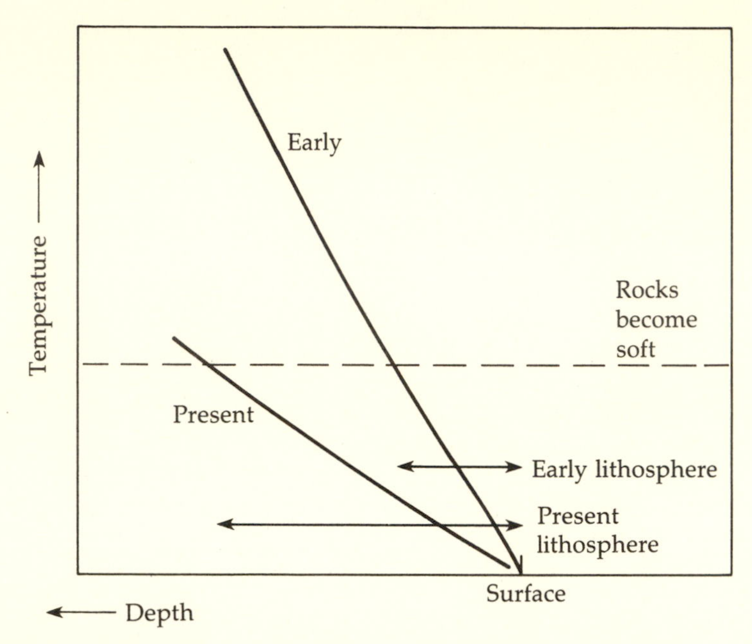

Figure 10–3. The effect of internal temperatures on the thickness of the lithosphere.

Higher internal temperatures should have resulted in a thinner, weaker lithosphere (Figure 10–3), which might, very plausibly, have tended to break up into smaller plates than those of today. Fróm a theoretical point of view the concept of microplate tectonics (Figure 10–4) on the early earth is attractive. What additional insight can be gained from the rock record?

Old rocks are preserved only in the interiors of the continental platforms. Because of sea-floor spreading and subduction, the rocks on the floor of the ocean are nowhere older than a few hundred million years. Oceanic rocks of greater age are of course preserved, but only where they have been incorporated into the continents as parts of mountain ranges. Still, only a small fraction of the surface of the globe has the potential of recording tectonic events of the distant past.

Careful geological mapping has revealed the remains of about 40 continental platforms that were formed more than 2 billion years ago (Figure 10–5). Their number and relatively small sizes are consistent with the idea of microplate tectonics. To-

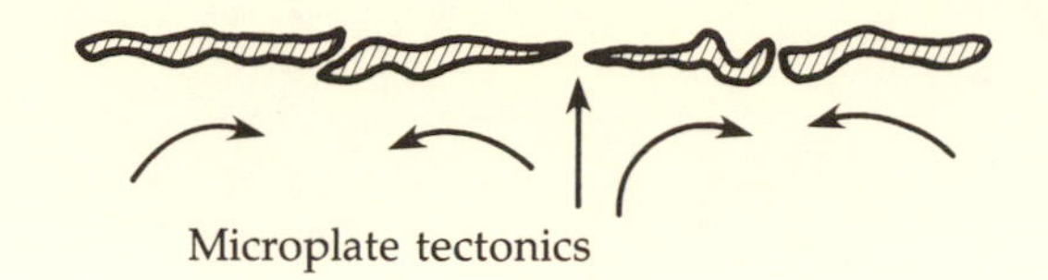

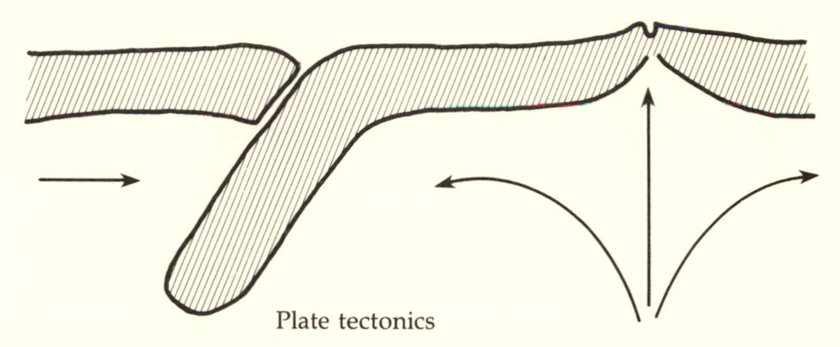

Figure 10–4. Hypothetical microplate tectonics compared with the plate tectonics of today.

gether these blocks of ancient rock make up about 5% of the continental crust. Their volume may have been much larger, of course, with only fragments having survived the ravages of erosion and later tectonic activity. Nonetheless, there is geochemical evidence for the continual creation of new continental crust at times extending almost to the end of the Precambrian. New continental crust, here, means granitic material that is released from the mantle rather than material formed by the melting of preexisting continental rocks. The latter process is a regular feature of mountain building. The geochemical evidence consists of detailed data on the concentration of minor elements in the new rocks and particularly on the concentrations of radioactive elements and their decay products that are signatures of either a mantle or a crustal origin. Unlike many new results in earth science, these geochemical conclusions have not generated strong controversy. It seems fairly clear that the total volume of granitic rocks in the crust of the earth increased with time during most of the Precambrian era.

Although continents could hardly have appeared on Earth

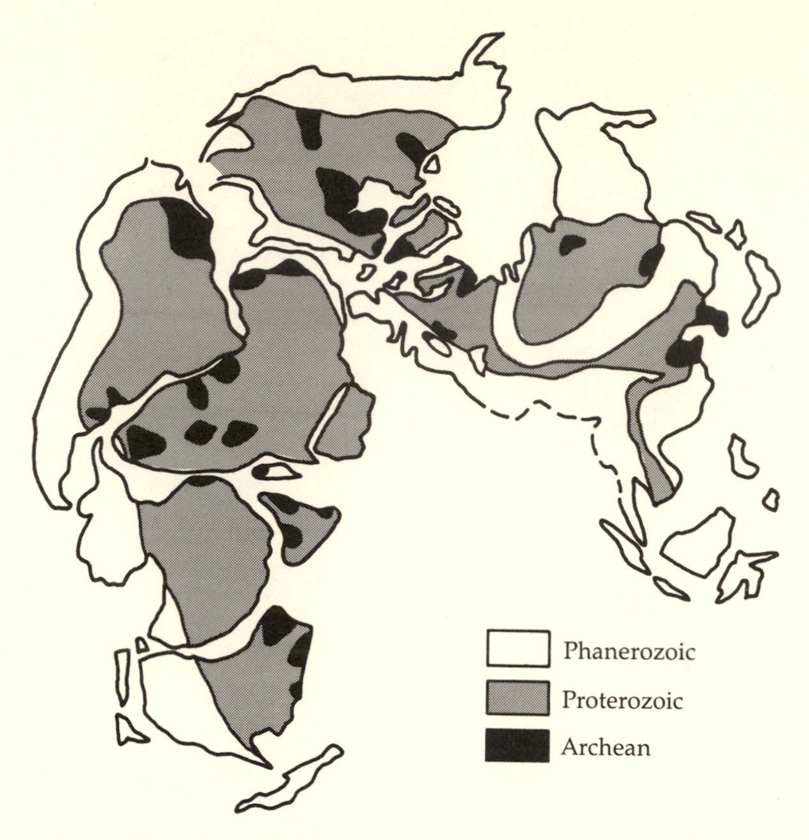

Figure 10–5. Remains of Archean continents are small and numerous. (Adapted from B. F. Windley, *The Evolving Continents,* Figure 1.1. John Wiley and Sons, New York, 1977.)

before the granitic material of which they are made was released from the mantle, it is important to distinguish between the growth of the total volume of granitic rock and the growth of continental land masses. The granitic rocks may, at one time, have been spread too thinly over the globe to build platforms rising above sea level. Growth of continents requires both a supply of granitic material and a mountain-building mechanism like plate tectonics to move the granitic material horizontally and pile it up. As a rule, igneous rocks do not show whether or not they formed in the interior of a continent, but the sedimentary rock record does provide evidence concerning the growth of continental platforms.

The composition of ancient sediments, particularly the chemical sediments deposited from solution, preserves indirect information on the composition of the ancient ocean. The data must be analyzed with care because of the possibility of more recent contamination, but Jan Veizer of the University of Ottawa has been able to recover a convincing record of subtle changes in ocean chemistry during the course of the Precambrian era. From an analysis of the concentrations of elements present in trace amounts, principally in carbonate sediments, Veizer has discovered that the composition of the Archean ocean was dominated by interaction with basaltic rocks (the rocks of the ocean floor). Today, ocean chemistry also depends on reaction with continental rocks. The data reveal an increase with time in the continental influence on the composition of sea water that was particularly rapid in the Early Proterozoic, between about 2.5 and 2 billion years ago. Because continents affect sea water composition principally through the weathering process, there is a strong indication that continental land masses first became extensive during this period of time.

Further evidence concerning the growth of continents comes from the study of the mineralogy, structure, and texture of sedimentary rocks. The record extends all the way back to 3.8 billion years ago at Isua, but the oldest well-preserved sedimentary rocks are found in the Barberton Mountain Land of South Africa and the Pilbara Block of Western Australia. These rocks date back to about 3.5 billion years ago. The very old formations lack platform deposits, horizontally stratified sediments deposited in shallow water over extensive areas of continental platform (Figure 10–6). The oldest platform deposits are found in the 2.9-

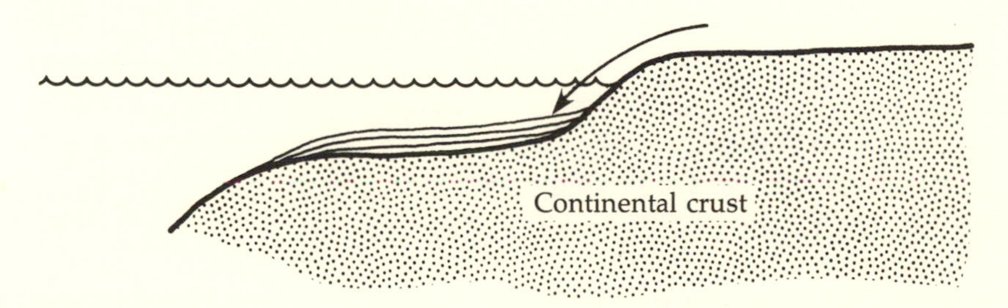

Figure 10–6. Platform sediments. The oldest ones date from 3 billion years ago.

billion-year-old Pongola Supergroup of South Africa and platform deposits become increasingly common in deposits of younger age. This observation in itself suggests that continental platforms were rare in the Early Archean.

Analysis of the oldest sedimentary rocks provides further confirmation of this picture. Deposits older than 3 billion years lack the granitic detritus that would have been present had they formed near continental land masses. Instead, these sediments are composed of volcanic debris that appears to have washed off the slopes of nearby volcanic islands. Indeed, the oldest sediments are invariably associated with an abundance of lava flows, many of which erupted under water. The combined deposit of lavas and sedimentary rocks is called a greenstone belt (Figure 10–7). The sedimentary detritus of the greenstone belts has suffered little mechanical abrasion or chemical change, which suggests that erosion and transport were rapid and the islands were small. Sediments deposited in deep water are found quite close to sediments deposited in shallow water, again suggestive of the steep slopes of a volcanic island. Great thicknesses of sediments deposited in deep water, such as those of the modern geosyncline, are not found. Formation of geosynclines requires a stable pattern of erosion and deposition over an extended period of time. Where the sedimentary deposits of greenstone belts are locally thick there is an alternation of deep and shallow water deposits that suggests intermittent sinking of a fairly small block of crust followed by rapid filling of the resultant basin by erosion of adjacent uplifted blocks (Figure 10–8).

Nearly all surviving sediments of Archean age are found in greenstone belts. Sediments younger than about 3 billion years, however, show evidence of an increasing influence of nearby blocks of continental crust. Sediments of Proterozoic age look very much like younger sediments, being composed principally of the products of weathering and erosion of continental material, and frequently having been deposited on stable continental platforms. Continents were evidently extensive in the Proterozoic but rare in the Archean.

Most greenstone belts are of Archean age. Their disappearance from the record by the Middle Proterozoic suggests a fundamental change in tectonic style during the Precambrian. Unfortunately there is still no agreement on the tectonic setting or

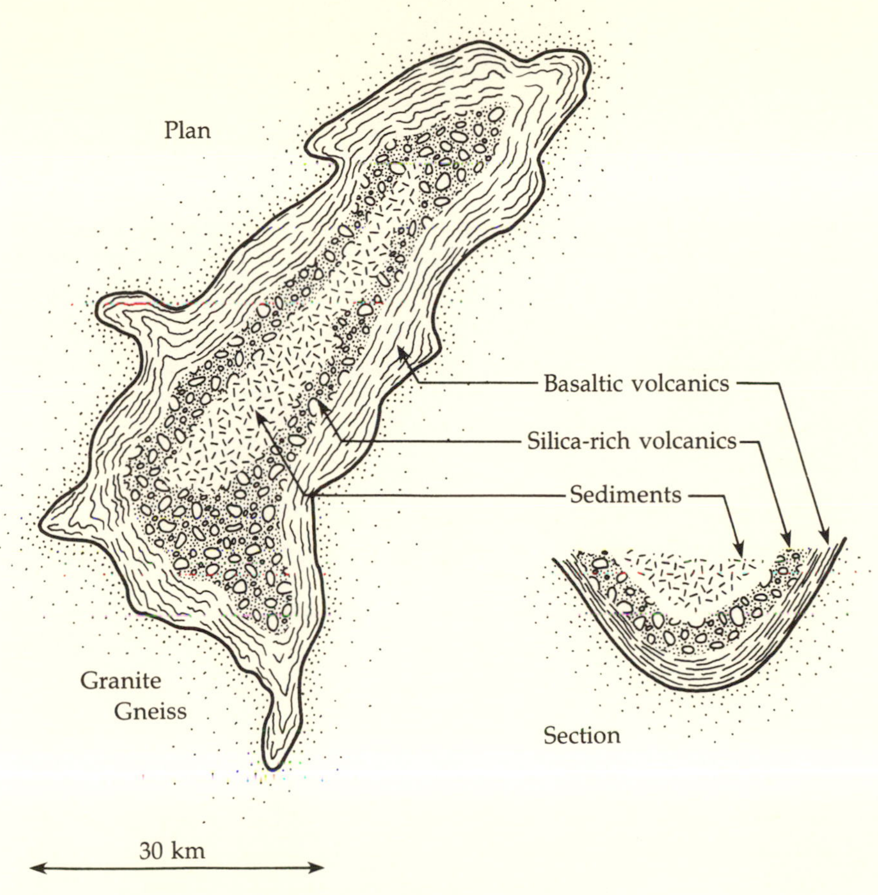

Figure 10–7. A typical greenstone belt.

significance of greenstone belts, so it is not clear what this change was.

A tentative summary of the Precambrian history of the crust would be as follows (Figure 10–9): During the first 700 million years or so of earth history the planet was in the last stages of accretion and the surface was subject to heavy bombardment by extraterrestrial material. No crust dating from this period of time, called the Hadean, has yet been discovered. Granitic rocks released from the mantle have been present at least since 3.8 billion years ago, but there is no evidence that the oldest of these had already built up into extensive continental platforms. The volume of granitic material increased throughout the Precambrian era,

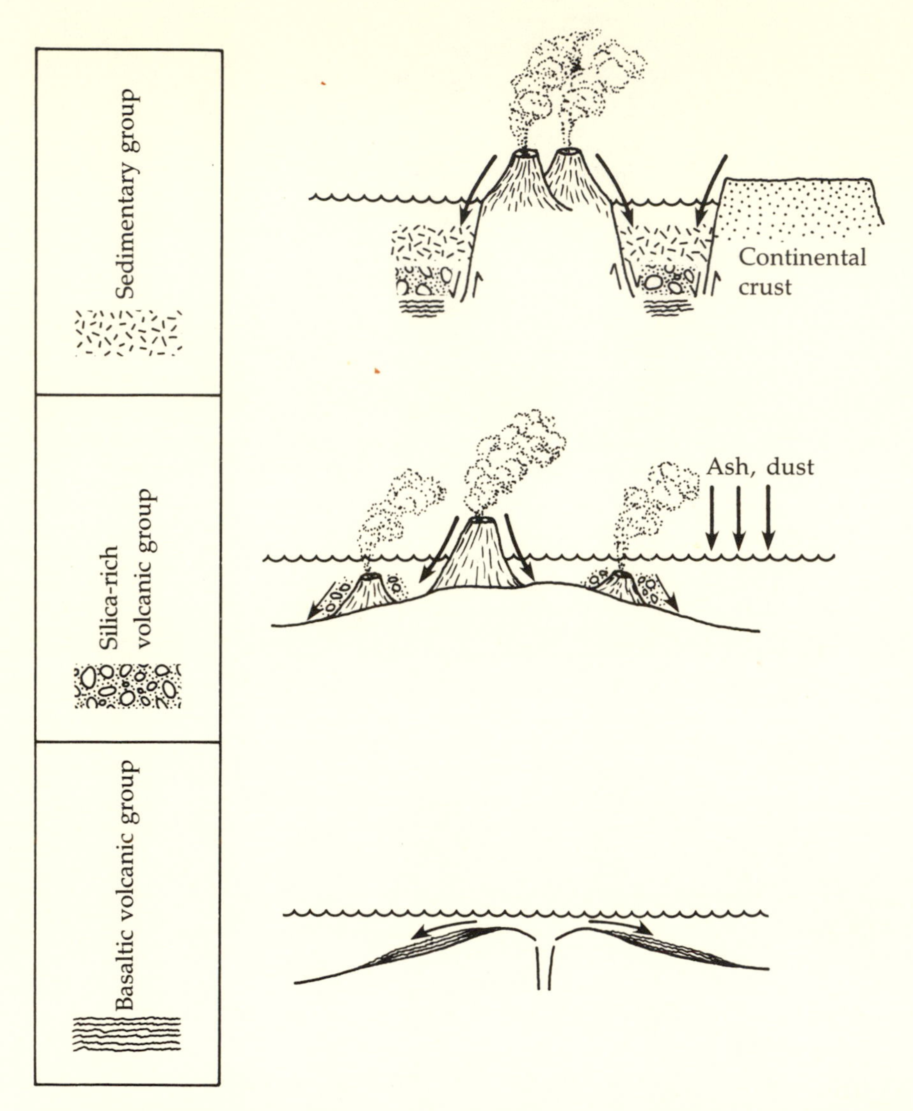

Figure 10–8. Interpretation of the environments of deposition of the sequence of rocks in a typical greenstone belt. The oldest rocks are at the bottom.

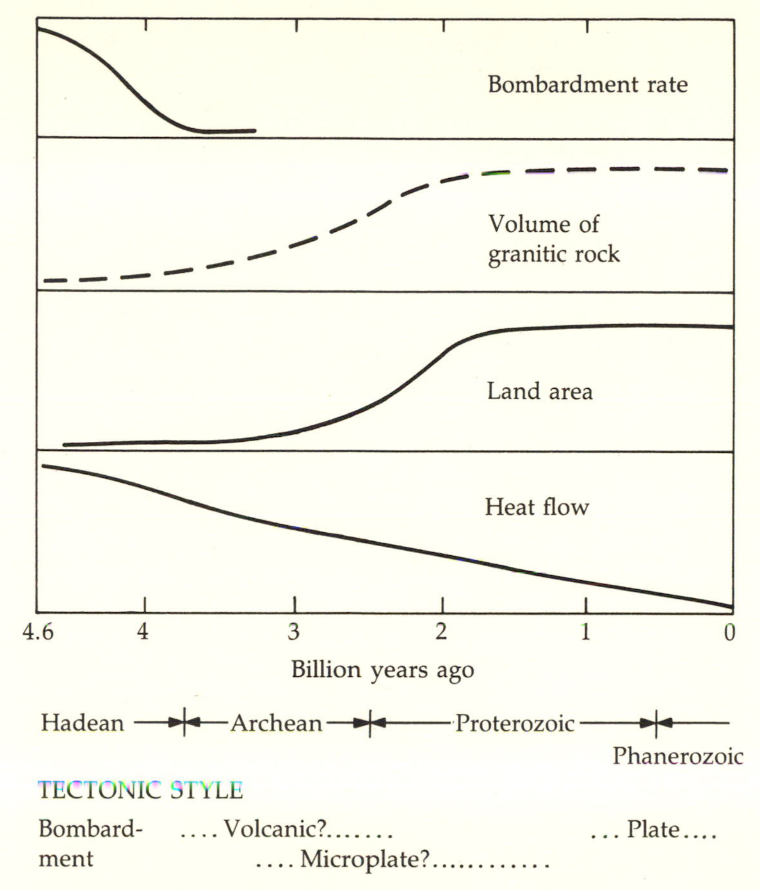

Figure 10–9. Possible evolution of terrestrial tectonics.

but the rate of this increase is not known. The first evidence of continental land masses dates from about 3 billion years ago, and continents apparently developed particularly rapidly around the end of the Archean and the beginning of the Proterozoic. The reason for this apparent spurt of continental emergence is not clear. Perhaps temperature gradients within the earth were too great during the Archean to permit continental thicknesses of granitic rocks to accumulate without melting at the bottom. Alternatively, perhaps the tectonics of the era did not include the horizontal sweeping action of plate tectonics that concentrates granitic material into thick, continental piles. For whatever reason, stable areas of continental crust were rare during the Archean and abundant during the Proterozoic. Such continents as did come into existence during the Archean were almost certainly small. Much of the Archean earth must have been covered by water, punctuated by ephemeral volcanic islands.

Very little is known about the environmental implications of these apparent changes in the face of the globe. Ocean chemistry has changed as a result of the increasing influence of continental material, but whether the changes have been profound enough to affect the course of biological evolution is not yet clear. Atmospheric composition and climate may well have been influenced also. The existing record of microfossils shows that habitable, shallow water environments have been present since at least 3.5 billion years ago, but they were probably restricted in area during the Archean and possibly short-lived in terms of geological time scales. The growth of extensive areas of stable continental shelf around the end of the Archean may have had a major impact on life, permitting a significant increase in biological productivity.

Suggested Reading

Tankard, A. J., M. P. A. Jackson, K. A. Ericksson, D. K. Hobday, D. R. Hunter, and W. E. L. Minter. *Crustal Evolution of Southern Africa*. New York: Springer-Verlag, 1982.

Windley, B. F., Ed. *The Early History of the Earth*. London: John Wiley and Sons, 1976.

CHAPTER 11

The Biosphere

ABUNDANT LIFE sets Earth distinctly apart from all the other planets in the Solar System. In simplest terms, this is true because of the impact of life on its physical and chemical environment. Aspects of this impact have been explored in earlier chapters, but the really fundamental distinction comes from the complexity of life itself. Life has created a diversity of forms and processes out of a small number of chemical building blocks that far exceeds what could be achieved by nonbiological means. Only in living organisms can a small change on the molecular level of organization (in the gene) cause a large and enduring change in the macroscopic form and function of the end product of growth and development. The result is that earth history is overwhelmingly richer than any history that can be imagined for a lifeless planet. Much more has happened on Earth, and most of it has involved the biota (the combination of all living organisms). In this chapter, I shall describe some of the features of the biosphere that are important to an understanding of earth history.

The biosphere is defined as the biota together with the part of the earth in which they live. Its extent includes all of the land surface (except for the highest mountains and the hottest, driest deserts), as well as the oceans and the top few meters of sea-floor sediments. It also includes the lowest few kilometers of the atmosphere. The name was first used by the Austrian geologist, Eduard Suess, in a short book on the Alps published in 1875, but the concept of the biosphere had little impact on scientific thinking before it was developed in the 1920s by the Russian mineralogist, Vladimir Ivanovitch Vernadsky. Earth history is richer than the histories of other planets because it embraces the history of the biosphere.

What properties of earth have made it the home for life? According to G. Evelyn Hutchinson of Yale University, a pioneer in the scientific study of the biosphere, there are three attributes that make Earth's surface habitable. First, it is a region in which liquid water can exist in substantial quantity. This is important because all actively metabolizing organisms consist of systems of organic molecules dispersed in aqueous media. Second, it offers an abundant external source of energy—sunlight—to support biological activity. Third, there exist, within the biosphere, interfaces between solid, liquid, and gaseous states of matter. Life is not impossible in the absence of interfaces, as evidenced by the freely floating plankton of the open ocean, but the concentration of organisms on the land surface and at the bottom of lakes and seas suggests that it is much easier to live at an interface, preferably one between solid and fluid.

To this list I would add a fourth essential feature of the biosphere: It contains the chemical elements on which life depends, and the supply of these elements is regularly renewed. What are these elements important to life, and what are the means of renewal? The answer to the first part of the question is that nearly all of the chemical elements are important to some organism or other. The number of elements that are essential to all living creatures is smaller, but still substantial. Of these I will select just a few key elements to illustrate the basic principles of biogeochemistry.

The most abundant elements in living organisms are hydrogen, oxygen, and carbon. These are readily available in the biosphere in the form of water and carbon dioxide. The 20 amino acids of which proteins are composed also contain nitrogen, and some contain sulfur. Nitrogen is less readily available to organisms, for reasons I shall describe below. Phosphorus is an essential nutrient element, and for several reasons it is quite scarce in the biosphere. As an illustration of the kinds of processes that affect the habitability of the biosphere it is worth asking what these reasons are.

First, little phosphorus is produced by the nuclear reactions in stars that form all of the chemical elements from hydrogen. Its abundance in the universe is almost 100 times less than those of silicon and sulfur, close neighbors of phosphorus in terms of atomic weight. Second, chemical theory suggests that much of the earth's phosphorus was locked up in the core during the process of core formation, in combination with iron and nickel. Third, the compounds of phosphorus that are most common in the biosphere are only slightly soluble in water. The result of this combination of factors is that organisms do not encounter much readily available (dissolved) phosphorus. Instead, phosphorus is the nutrient element that limits the multiplication and growth of organisms in many areas of the ocean today (see Chapter 2). Iron is another nutrient element that is frequently limiting in the open ocean. It plays an essential role in the mechanisms of metabolism associated with oxidation and reduction reactions. Although abundant in the lithosphere, the insolubility of the oxidized form makes dissolved iron exceedingly rare in open water today. The Archean ocean, however, lacked oxygen and was presumably rich in iron (see Chapter 4). A shortage of dissolved iron was therefore not a problem for the first organisms.

The solubilities of a number of other nutrient elements depend, like iron, on their states of oxidation, which suggests significant changes in the relative availability of nutrients within the biosphere at the time of the rise of atmospheric oxygen. Metabolic pathways relying on formerly abundant nutrient elements may have lost their competitive advantage to new pathways that took advantage of newly available nutrients. The impact on the subsequent course of biological evolution may have been large, but no scientists with the necessary combination of biochemical and geochemical expertise have yet addressed the question. Although the hypothesis that the availability of nutrients changed significantly when oxygen first became abundant is entirely speculative, we have here a good example of the potential interactions of biota and environment, with ramifications extending on through time, that are so much a feature of earth history.

Biogeochemistry is the subject concerned with the chemical interactions between living and inanimate parts of the biosphere. It examines the processes that maintain the supply of nutrients within the biosphere as well as the impact of the biota on the chemical composition of the biosphere.

The clearest example of this impact is the abundance of oxygen in the present-day biosphere. The processes by which organisms maintain this bubble of oxygen in the midst of a universe composed overwhelmingly of hydrogen and other elements that react with oxygen were described in Chapter 8. Other examples abound. As I mentioned in Chapter 4, life influences the carbon dioxide content of the atmosphere, particularly in terms of geologically short time scales. One major environmental concern today is the increase in carbon dioxide that is resulting from the burning of fossil fuels. The possible effect of human activities on another atmospheric trace constituent, ozone, has been much in the news in recent years. Many of the trace constituents of the atmosphere, including hydrogen, methane, nitrous oxide, carbon monoxide, ammonia, hydrogen sulfide, and sulfur dioxide, are biogenic waste products.

The impact of the biota on the composition of the environment is only one aspect of biogeochemistry. The processes that maintain the supply of the nutrient elements that sustain life are equally important. The question of supply demands attention because of the continual drain of material from the land into the sea and because of the inexorable tendency of solid particles, incorporating nutrient elements, to accumulate as sediments at the bottom of the sea, a process that removes nutrients from the biosphere. If the habitability of the biosphere is to be maintained, cyclic processes must return nutrients to the land. Water provides an example of a particularly simple geochemical cycle. Water runs off the land into the sea, but is returned to the land by the processes of evaporation, transport, and precipitation (Figure 11–1). If water were not recycled in this fashion, life on land would be impossible.

But what about nutrients such as phosphorus that do not evaporate? What mechanism restores them to the biosphere after they have been locked up in sediments on the bottom of the sea? What happens to the calcium and carbon drained out of the biosphere when carbonate sediments accumulate? The answer lies in tectonics. Most nutrient elements are recycled to the biosphere by tectonic activity (Figure 11–2). Without such activity the ocean basins would gradually fill with inaccessible debris, the land would erode away, and the supply of the chemical elements required by life would gradually taper off. The action of plate tec-

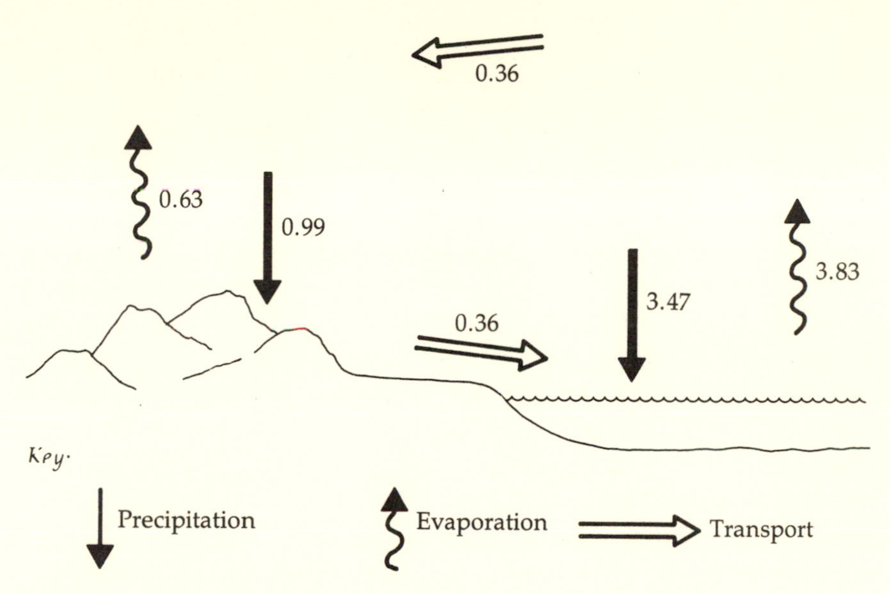

Figure 11–1. The cycle of water among ocean, atmosphere, and land. The numbers refer to amounts of average annual transfer in units of 10^{20} grams.

tonics carries sediments to the edges of the ocean basins, where they may contribute to the lavas and gases of the volcanoes over subduction zones or be lifted above the surface of the sea in the form of mountain ranges composed of folded sedimentary rocks. To a large extent, the fertility of the biosphere is sustained by the recycling of nutrients that results from tectonic activity and

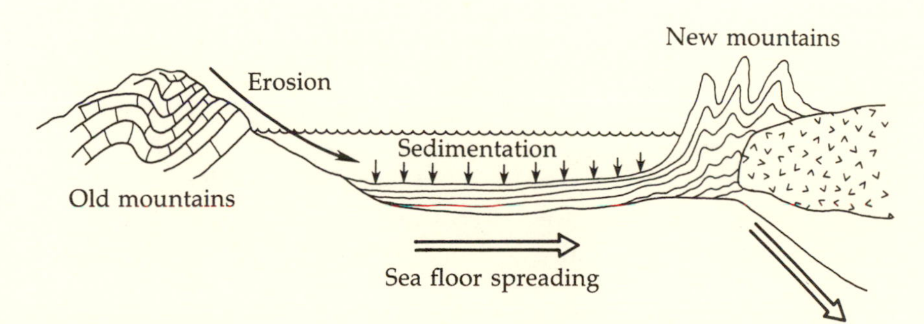

Figure 11–2. Tectonic activity sweeps debris out of the ocean basins, returning sediments to the land from which they were derived.

mountain building. Abundant life may not be possible on a planet that lacks a suitable tectonic style. As I noted in Chapter 9, none of the other planets appears to have the plate tectonic style of Earth. Maybe they could not sustain biospheres even if their temperatures were more comfortable. The early earth was tectonically active, although the style may have differed from that of the present day (Chapter 10). In all likelihood, terrestrial tectonics have always provided for the geochemical recycling of elements essential to the biota.

But this is not the whole of the story of nutrient supply. Although large-scale geochemical cycles sustain the fertility of the biosphere, a high level of biological activity is made possible only by rapid biological cycles that transfer nutrients from biota to the environment as fast as organisms extract them (Figure 11–3). Processes of decay, frequently accelerated by the metabolic activities of microbes, are important in this connection. By preventing a steady growth in the total mass of organic debris, they restore essential nutrients to the biosphere (Figure 11–4). The most striking example of a rapid biological cycle is the one that cycles carbon and oxygen between biota and environment via the processes of photosynthesis and respiration (discussed in Chapter 1). A stable biosphere would not be possible in the absence of either one of these metabolic processes.

Nitrogen offers a further illustration of how biological processes recycle essential nutrients through the biosphere. It might at first glance appear that the most abundant constituent of the atmosphere is of all nutrient elements the one least likely to be in short supply. This is far from the case because nitrogen gas is chemically unreactive; most organisms cannot metabolize it. Instead they depend on a few species of prokaryotic microbes for a supply of what is called fixed (or combined) nitrogen, nitrogen atoms combined with either hydrogen or oxygen to form ammonium, nitrate, or nitrite ions (Figure 11–5). Organisms able to fix nitrogen gas presumably evolved early in the microbial era of earth history when the abiotic source of combined nitrogen became inadequate to sustain further expansion of life (see Chapter 6). But all of Earth's nitrogen could long since have been extracted from the atmosphere if the biological nitrogen cycle had not been completed by the evolution of denitrifying bacteria, which derive energy from nitrate and nitrite by converting them back into ni-

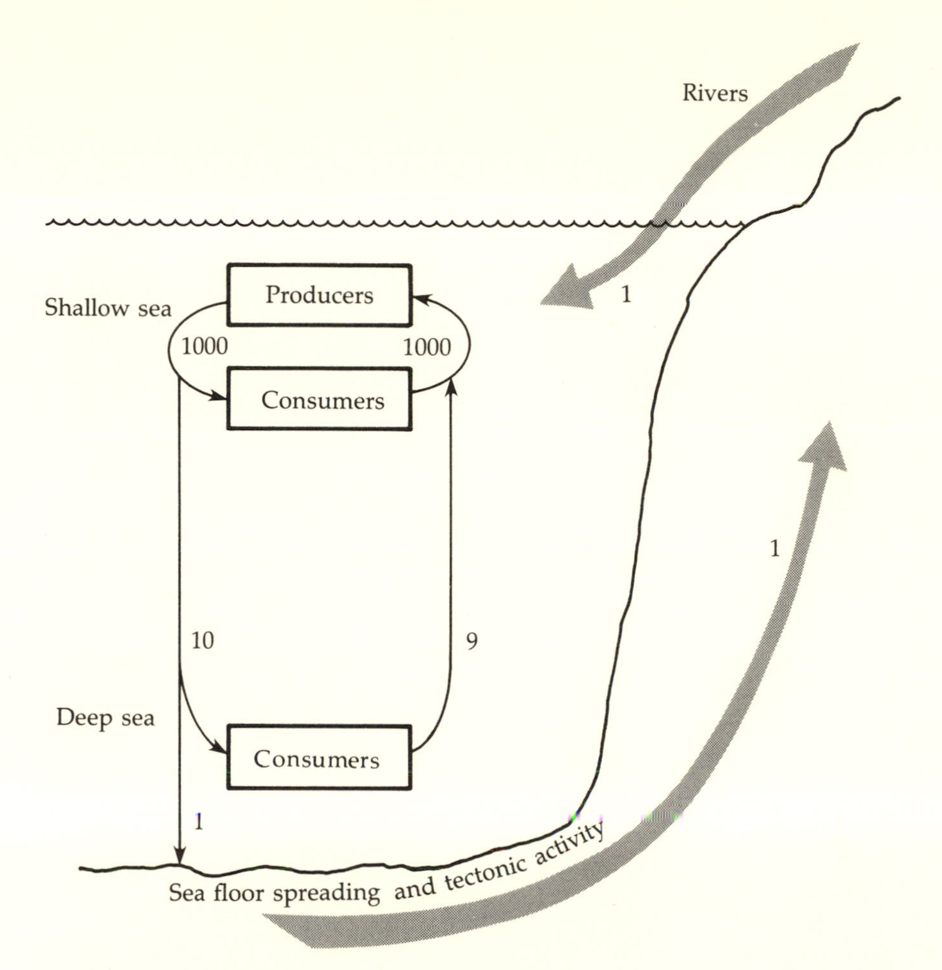

Figure 11–3. Biological recycling. The productivity of the biosphere depends on rapid recycling by organisms of essential nutrient elements. Shown here are relative fluxes of phosphorus in the ocean.

trogen gas, as well as nitrifying bacteria, which use the oxidation of ammonium ions to nitrate and nitrite as their energy source.

All of these organisms act in concert to maintain a steady flow of biospheric nitrogen through its various chemical forms and through the biota. The rate of nitrogen fixation is restrained by its expense in terms of metabolic energy. Organisms fix nitrogen only if there is no combined nitrogen available to them. For this reason, the ratio of fixed nitrogen to phosphorus is almost

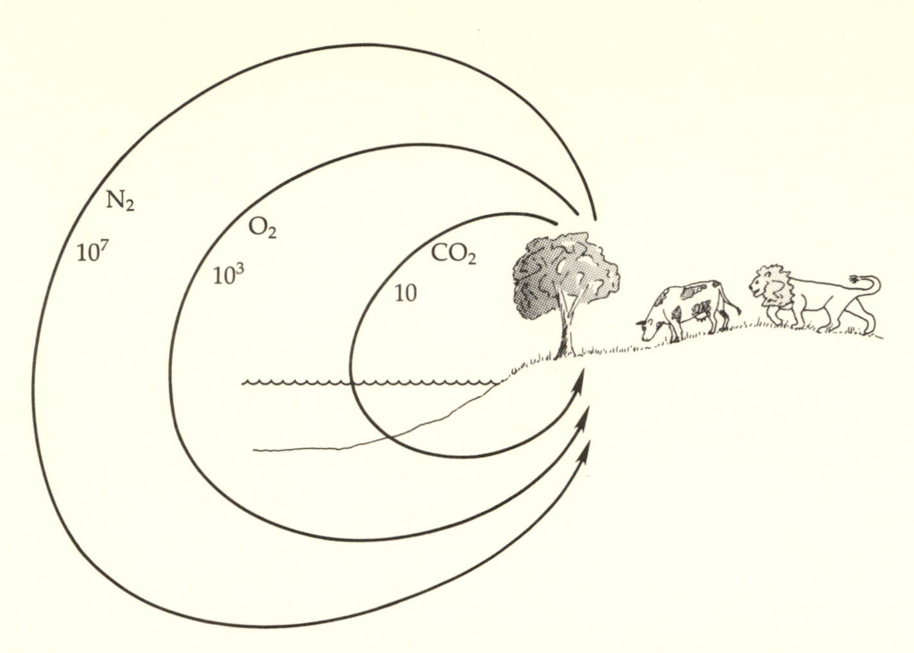

Figure 11–4. Biological cycles of the major atmospheric gases. The numbers give the average time in years for a molecule to travel once around the cycle from biota to atmosphere and back to biota.

exactly the same in sea water as in organisms. Oceanic microbes fix nitrogen if a shortage of nitrogen is limiting their growth but do not fix nitrogen if it is already more abundant, relative to their needs, than phosphorus. It is not yet known how long the biogeochemical cycle of nitrogen has existed in its modern form. On the one hand, the suggestion made above, that nitrogen fixation developed early, is supported by the fact that this metabolic capability exists only in prokaryotic, anaerobic, presumably primitive, microbes. Nitrification, on the other hand, is an aerobic process—organisms convert ammonium ions to nitrate only in the presence of oxygen. Presumably the introduction of nitrification to the metabolic repertoire was delayed until after the oxygen revolution. Denitrification is an anaerobic process, but it requires nitrate. Today nitrate is the most abundant form of combined nitrogen in natural waters, but it may have been in short supply before the world became aerobic. I would guess that the anaerobic Archean ocean contained ammonium ions in place of

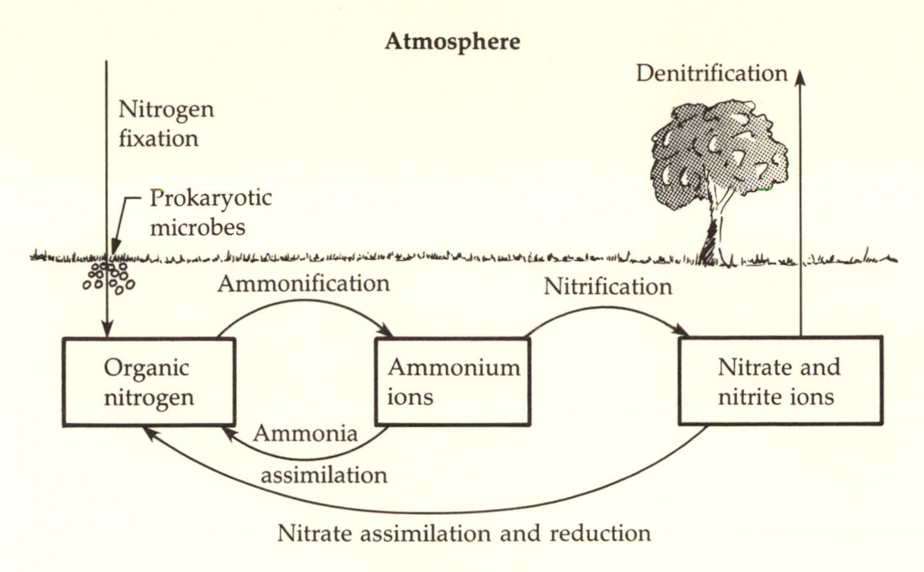

Figure 11–5. Cycles of nitrogen within the biosphere.

nitrate ions, that the change took place after the rise of oxygen as a consequence of the introduction of nitrification, and that denitrification was the most recent process of the biological nitrogen cycle to develop.

The nitrogen cycle, with its nitrogen fixers, nitrifyers, and denitrifyers, calls attention to the structure of living communities. Life as we know it is not possible without a variety of different organisms living in close association and interacting with one another. This is as true for metazoa and metaphyta as it is for microbes. In a mature and stable community there is an orderly flow of energy and nutrients through the various levels, called trophic levels, of a grazing chain (in reality more often a web than a chain, but the chain illustrates the principles). On the lowest trophic level (Figure 11–6) are the primary producers of organic matter, the autotrophs (usually plants). On the next level are herbivores, which eat the autotrophs. Then there are carnivores, which eat the herbivores, and possibly still another level occupied by larger carnivores that eat the smaller carnivores.

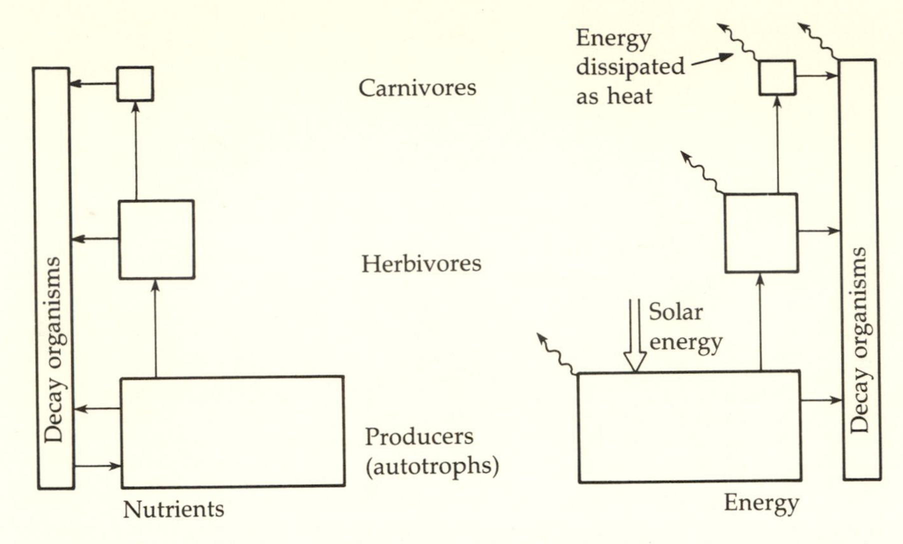

Figure 11–6. Flow of energy and nutrients through the grazing chain of a mature, natural community. The input of solar energy is ultimately dissipated as heat released during respiration, while nutrients are largely recycled.

Ecology is the science concerned with the interactions of organisms in natural communities. One of its most successful methodologies has been the study of the flow of energy through the different trophic levels of a community, an approach pioneered by Raymond L. Lindeman while he was working at Yale

LIMITING NUTRIENTS

According to Liebig's Law, organisms multiply and expand in number until they exhaust the supply of some resource essential to their growth. In the oceans the concentrations of dissolved phosphate and nitrate fall to small values in the surface waters where phytoplankton, exposed to sunlight, are growing (Figure 11–7). Phosphorus and nitrogen are essential nutrient elements. The plankton extract these elements from the water until their concentrations are too low to support further growth. Growth is therefore limited by the rate of supply of phosphorus and nitrogen (and possibly other elements as well) determined by upwelling and mixing of deep water rich in nutrients.

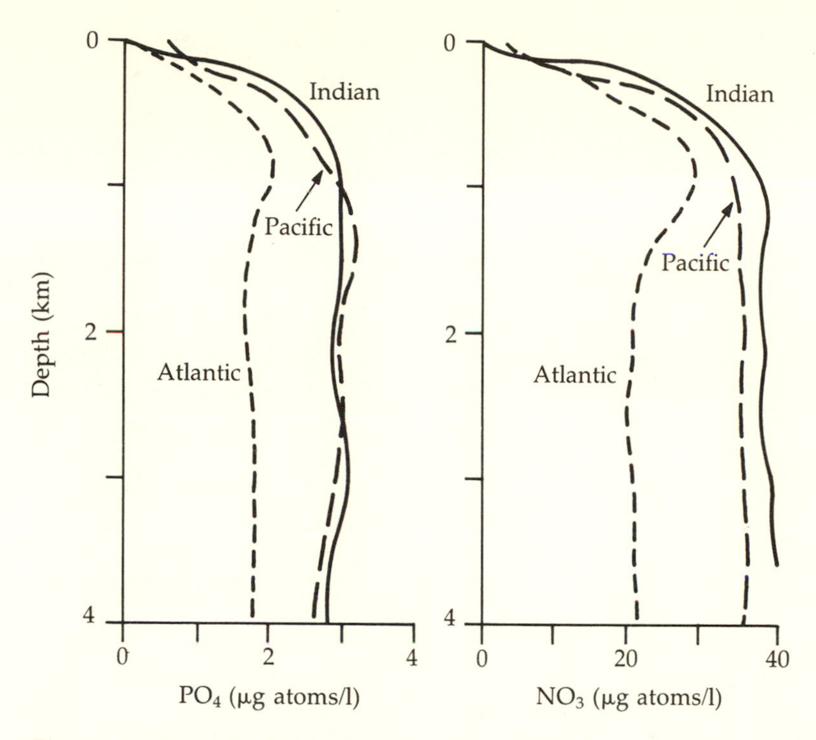

Figure 11–7. Vertical distributions of phosphate and nitrate in the Atlantic, Pacific, and Indian Oceans. (Adapted from K.K. Turekian, *Oceans,* p. 90. Prentice-Hall, Inc., Englewood Cliffs, NJ, 1968.)

University with G. Evelyn Hutchinson in 1942. Because the organisms at each trophic level must use a substantial fraction of the energy available to them simply to maintain themselves without growth, available energy decreases markedly with each step up the grazing chain. For this reason, plants are more abundant than herbivores, which in turn are more abundant than carnivores, and so on. A series of field and laboratory studies stimulated by Lindeman's suggestion has led to the formulation of the so-called "Ten Percent Law." This law states that in nature some fraction of the energy entering any population is available for transfer to the populations that feed on it without serious disruption of either prey or predator populations. A rough average value of this fraction appears to be between 10% and 20%.

Paleoecology is concerned with the evolution and history of this and other ecological relationships and with the ecological interactions of past communities of organisms. Biological evolution has been heavily influenced by the struggle to eat or to avoid being eaten. According to the Ten Percent Law, predators have always been fewer in number (or at least less in total mass) than their prey. In many cases their necessarily small populations have rendered them relatively vulnerable to extinction. Moreover, the grazing chain cannot extend to ever higher levels. The population of a large predator (call it a dragon) that ate lions and tigers, for example, would be so small that dragons would seldom meet. Therefore dragons do not exist. Paleoecological studies offer the means by which we can hope to understand the diverse and complex history of biological evolution, the subject of later chapters.

The grazing chain may be a fairly late development in terms of geologic history. Predation was probably not a major selective pressure in biological evolution until after the origin of the eukaryotic cell, perhaps 1.4 billion years ago. Much of Precambrian time was taken up by the evolution of metabolism. Life's original struggles were with the nonliving environment—how to metabolize available nutrients, including the waste products of other microbes, and how to defend against potential poisons like oxygen. Organisms interacted, but most of their interactions were passive and chemical in nature. The eukaryotic cell, in contrast, could be large and mobile. Some eukaryotes achieved a richer diet by ingesting whole cells of smaller microbes. Thus the stage was set for the evolution of multicellular creatures of diverse morphology and behavior, and much of this diversity was directed toward more efficient predation or more effective protection.

Suggested Reading

Bowen, H. J. M. *Environmental Chemistry of the Elements.* London: Academic Press, 1979.

Colinvaux, P. *Why Big Fierce Animals Are Rare.* Princeton, NJ: Princeton University Press, 1978.

Hutchinson, G. E. *The Ecological Theatre and the Evolutionary Play.* New Haven: Yale University Press, 1965.

Scientific American. *The Biosphere.* San Francisco: W. H. Freeman and Co., 1977.

PART THREE *Mankind*

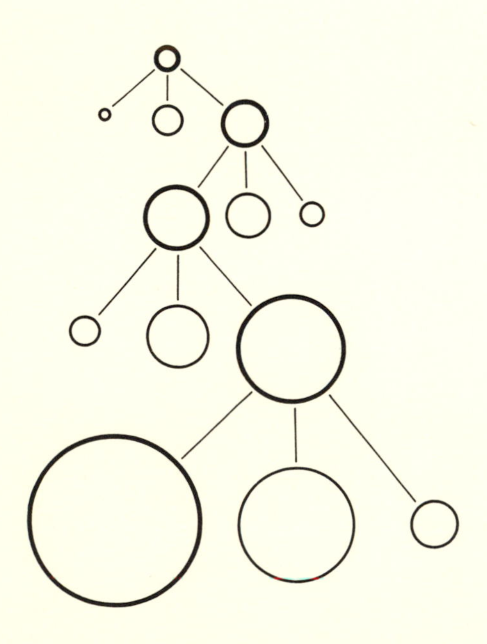

CHAPTER 12

The Age of Animals and Plants

THE MOST RECENT 570-million-year era of earth history is called the Phanerozoic—"the age of obvious animals"—in allusion to its abundant fossil record. In contrast, the Precambrian era (with which we dealt in earlier chapters) was at one time called the Cryptozoic ("crypto" means "hidden or secret"). The emphasis on animals implied by the name does not mean that animals have somehow been more important than plants or microbes during recent earth history. Animals and plants are entirely dependent on one another. Plants convert carbon dioxide into organic matter that animals can eat. Animals convert organic matter back into the carbon dioxide needed by plants. The geochemical role of microbes has not diminished during the Phanerozoic either. Microbes are still largely responsible for determining the chemical composition of our environment, just as they were during the Precambrian. The emphasis on animals implied by the term Phanerozoic, as well as its subdivisions—Paleozoic, Mesozoic, and Cenozoic (see Figure 1–1)—simply reflects the fact that the fossil record of the evolution of animals is more conspicuous than the record of other major groups of organisms.

Animals dominate the fossil record because many of them produce mineralized organs such as bones, shells, and teeth that are almost as sturdy as rocks. These organs have a relatively high potential for preservation. Very few plants have mineralized organs, but plants have responded to the struggle for survival by evolving a number of organic compounds that are resistant to bacterial or fungal decay and are fairly common in the fossil record. Durable plant organs, rich in these compounds, include wood, seeds and fruits, spores and pollen, and the waxy cuticle that coats leaves and stems. Even the less resistant parts of plants, such as flowers, have been preserved under favorable circumstances when their cellular material has been replaced by dissolved minerals. This process of petrifaction was also responsible for the microfossil record of the evolution of microbial life that has already been discussed.

Soft parts of both animals and plants are occasionally preserved as molds or casts, surface impressions that formed when a fine-grained sediment accumulated and hardened around the part before it decayed. Tracks left in mud by creeping or burrowing animals are another source of information. The earliest preserved tracks providing undisputed evidence of multicelled animals (metazoa) date from just before the end of the Precambrian.

The major evolutionary developments during most of the Precambrian involved metabolism, the processes by which organisms extract energy and cell building material from the environment. Changes in structure and form did occur within the cell, but most Precambrian organisms would look much the same to the untutored eye. Metabolic developments during the Phanerozoic, however, have been less important. Evolutionary experimentation has mostly affected form, structure, and organization, and has brought about an increasing variety and complexity of strategies for exploiting the natural economy. The passage of time has led to the development of ever more complex multicelled creatures, with internal organs that are more highly differentiated and specialized, and structures that are increasingly elaborate. Older, less highly evolved forms of life have, of course, survived; prokaryotic microbes are an example. Their modern representatives are the key to an understanding of the fossils of their ancestors.

The rich fossil record of the Phanerozoic era poses a number of fundamental questions. First, what essential biological innovations have permitted the diverse and complex biota of today to evolve from Proterozoic microbes? Second, what mechanism of change has permitted new species to emerge? Third, what factors have determined the direction of evolution? Fourth, what factors have controlled the rate? Scientific understanding of evolution is sufficient to provide partial answers to these questions, but only in very general terms.

The basic ideas of Charles Darwin and Alfred Russel Wallace concerning the origin of species by natural selection are common knowledge today. A species is a group of individuals who share a pool of common genes and are potentially able to interbreed. They are not, however, genetically identical to one another. Novel genes arise as the result of mutation, and novel combinations of genes result from the sexual method of reproduction, the recombination of genetic material from each parent in the germ cells of the offspring. The genetic differences result in heritable morphological differences among the individuals of a species; some are bigger than others, for example. Because organisms can produce many more offspring than available resources can support, not all individuals compete effectively or survive long enough to reproduce. The ones that do are more often the ones with differences that provide an advantage in the struggle for survival, the ones with the higher fecundities. Thus genes that enhance the chances of reproductive success tend to become widespread in a population, while the genes that diminish this probability tend to disappear (Figure 12–1).

The genetic composition of an actively interbreeding population can change with time as a result of these processes, generally in a direction that adapts the population more closely to its environment. If populations of individuals of the same species are isolated from one another, by geographic barriers, for example, their genetic compositions can, and do, gradually diverge. Divergence results from the multitude of possible adaptive strategies, particularly in response to environmental differences among domains. When sufficient genetic difference has accumulated, divergent populations lose the potential to interbreed. After this has happened they are, by definition, no longer members of the same species. Thus new species can, in principle, emerge as a result of natural selection of individuals coupled with reproductive isolation of populations (Figure 12–2).

Processes that can lead to the emergence of new species are currently under active investigation. It is not clear that the process described by this classical theory has been of major importance in earth history. The problem is how to assess the relative importance of gradual adaptation to a new environment compared with

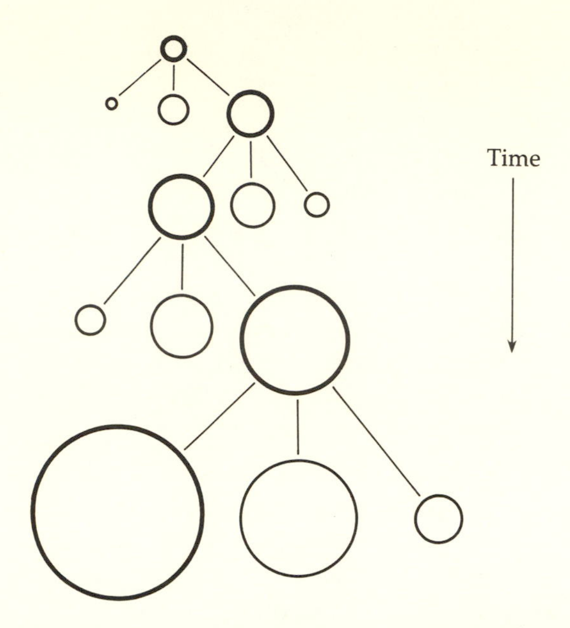

Figure 12–1. Evolution by natural selection. Suppose that genetic variability causes differences in the sizes of offspring and suppose also that larger individuals have more chance of surviving and reproducing. In time the average size of the individuals in the population will increase.

random changes in genetic composition in different species. The possibility of random change can be most easily understood in terms of the "founder effect" (Figure 12–3). Suppose that a small population of a widespread species becomes geographically isolated. The gene pool of these founders may differ in composition from that of the rest of the species simply by chance. In human terms, for example, none of the founders may carry the gene for blue eyes. Inbreeding in the small population may lead to further random loss of particular genes. If, in addition, competitive pressures are relaxed in the isolated environment, a rapid increase in numbers may occur. Selection pressures against deviant individuals may be relaxed during the course of this "population flush," further favoring the success of new and previously improbable combinations of genes. So new species can evolve rapidly as a result of the isolation of small populations, and this evolution can be a consequence of essentially random elements in the genetic composition of the founder population and of the essentially random reproductive history of the early generations

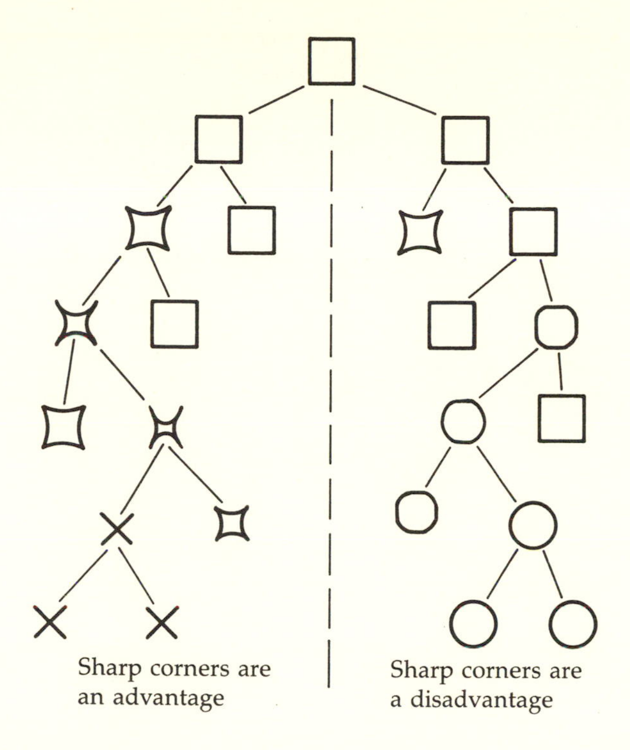

Figure 12–2. Speciation can result from different selective pressures on isolated populations.

of this population, rather than a result of the gradual adaptation of the population to its new environment.

This kind of chance evolution of new species appears to have occurred among the fruit flies (*Drosophila*) of Hawaii. Nearly a third of all *Drosophila* species yet described are found only in Hawaii, and many of these species are endemic to particular islands. The evidence suggests that there have been a small number of interisland colonization events as flies were blown from one island to the next. Single fertilized females may have been responsible for some of these colonizations. Laboratory studies of small populations of fruit flies have also revealed new species, reproductively isolated from their source populations, after only a few generations of separate evolution. New species of fruit flies can, however, also develop in laboratory populations by gradual adaptation to different environmental conditions such as temperature or humidity.

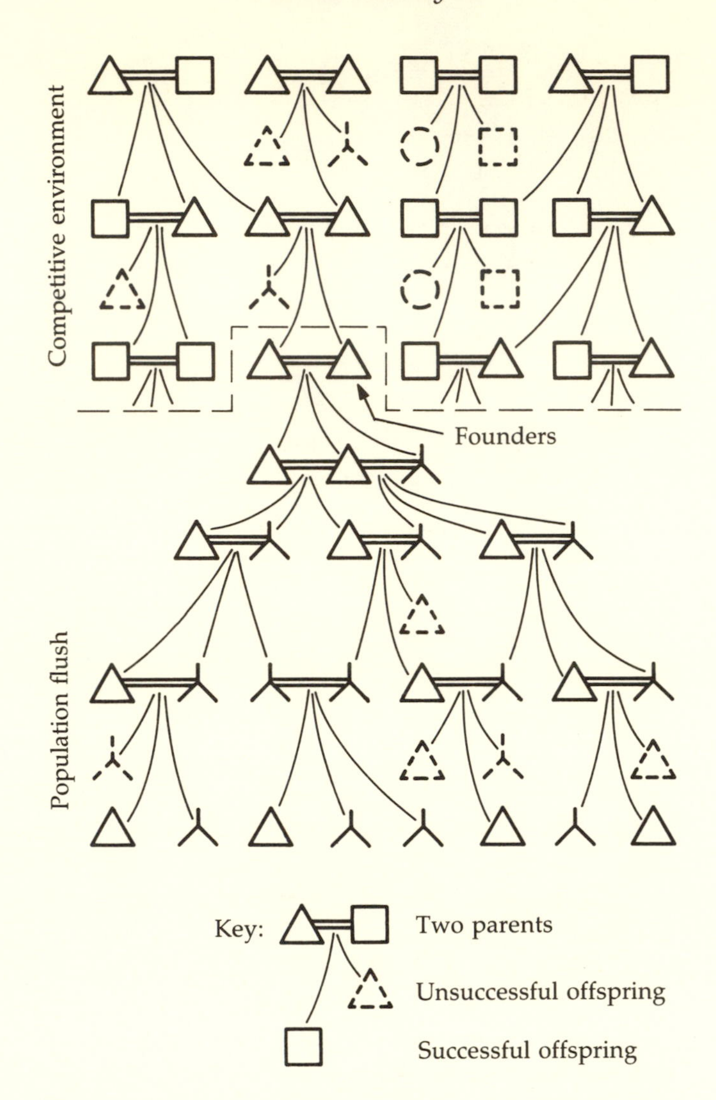

Figure 12–3. The "founder effect" and the "population flush." In the competitive environment at the top, about one-third of the individuals in each generation fail to reproduce. In particular, ○ and ⅄ are always unsuccessful. In the newly colonized environment at the bottom there are no genes for □ and ○. Reduced competitive pressure enables ⅄ to survive and reproduce.

Paleontologists have found evidence in the fossil record for both the gradual divergence of new species from old and the sudden appearance of new, exotic species. It is probable that gradual evolution in response to natural selection and rapid more or less random speciation have both occurred during the course of earth history. Their relative importance to the development of the biota has not yet been determined. It is clear, however, that even new species that originate in a small, isolated population are subject to natural selection (of species rather than of individuals) once the population flush is over. If they are to contribute to the history of life they must compete successfully with other species. One way to be successful is to occupy previously vacant ecological niches. The history of life has to a considerable extent, therefore, involved the progressive adaptation of organisms to new environments and the progressive evolution of new strategies for survival. I shall attempt to demonstrate this by examining major milestones in the evolution of life.

The fossil record can be examined from two disparate points of view. In this chapter I shall concentrate on lineages and pathways of evolution—the questions of what organisms are ancestral to which groups of organisms and what competitive advantages were conferred by successive evolutionary developments. The alternative approach, which will be the focus of later chapters, concentrates on the over-all structure of the biota, particularly on the diversity of life (numbers of distinct species), how it has varied with time, and possible causes of these variations.

Because plants, as a source of food, must necessarily precede animals in the opening up of new territory, I shall discuss them first. The history of plants as primary producers of organic matter begins with the blue–green bacteria. Microfossils of organisms that may well have been blue–green bacteria have been discovered in rocks 3.5 billion years old in Western Australia. Blue–green bacteria use the energy of sunlight to convert carbon dioxide into cell material. They are prokaryotic microbes, having small cells that lack a nucleus. They reproduce by simple cell division in which the parent cell divides into two genetically identical offspring cells. There are mechanisms by which such bacteria can exchange genetic information between cells, but this exchange is haphazard and not a requirement for reproduction. Surviving blue–green bacteria, and presumably their earliest ancestors, are in some cases unicellular, with single cells living in isolation, and in some cases colonial, with clusters of identical cells living together, loosely attached to one another by a sticky sheath. The colonial microbes formed the bacterial mats that have been preserved in the rock record as stromatolites.

Probably the most important event in the history of plants and animals was the origin of the eukaryotic cell. This cell is much larger than the prokaryotic cell. It conducts many of its biochemical functions in specialized organelles within the cell and its genetic material is packaged within a nucleus in the form of chromosomes. The eukaryotic cell permits sexual reproduction in which genetic information from two parents is combined in the offspring. This method of reproduction provides for great genetic diversity in a population and facilitates biological evolution. All of the higher plants and all animals are composed of eukaryotic cells.

The fossil record contains fairly convincing evidence of eukaryotes as much as 1.4 billion years ago. The first eukaryotic plants may have resembled the modern algae. They were plants like kelp and sea lettuce that are restricted by fairly simple internal organizations and methods of reproduction to a life almost entirely in water. In spite of their relative simplicity, many algae do exhibit the important property of differentiation during the course of development. Differentiation means that different cells within the organism develop quite distinctive properties and functions, although they are all descended from a common ancestral cell. Thus, in algae some cells can specialize to hold the organism fast to rocks, other cells perform photosynthesis, while still other cells are active in reproduction. Significant differentiation is present only in eukaryotic organisms. A multicellular, differentiated organism, with distinct organs and cells specialized to perform particular functions, is a very great advance over the colonial organizations of the prokaryotes, in which many cells can be associated, but nearly all of them are identical and perform identical biochemical functions. (For a discussion of how cooperation among cells may have evolved, see *Early Life* by Lynn Margulis (Science Books International, 1982).)

Algae or primitive metaphyta (multicellular, differentiated plants) probably adapted to a fresh-water existence before making the transition to land. The problems of moving from an aquatic environment to a terrestrial one were severe. Mechanisms had to be developed to prevent desiccation, to furnish a sturdy supporting structure, and to provide for reproduction and dispersal. In spite of the difficulties, the land surface offers plants advantages over an aquatic environment. They can keep one end buried in the soil, where nutrients are generally abundant, and can spread their leaves over a large area at the other end to capture unattenuated sunlight for photosynthesis. Because the force of the wind is less than that of currents and waves, land plants can develop larger and more delicate structures for this purpose. Nevertheless, there are factors that limit the expansion of plants on land just as there are in the ocean. The growth of oceanic plants is generally limited by the availability of sunlight and inorganic nutrient elements, principally phosphorus and nitrogen. Plants cannot grow at depths to which sunlight does not penetrate. Near the surface, where sunlight is abundant, the population expands until the supply of biologically usable phosphorus or nitrogen is exhausted (see page 124, Figure 11–7). On land the limiting factors are most often light and water. Where water is abundant, as in a rain forest, the plant population grows until all the photosynthetically useful sunlight is absorbed. In deserts, at the other extreme, sunlight is wasted because there is not enough water to sustain many plants. Other factors that can limit the expansion of plant biomass, such as low temperatures, high winds, absence of soil or nutrients, presence of toxic elements in the soil, or grazing, are probably effective only over relatively restricted areas of the globe.

Any areas of land where plants are not using either all of the available sunlight or all of the available water therefore represent unfilled opportunities for expansion of the plant kingdom. Much of the history of land plants reflects their evolutionary response to these opportunities. Evolution in the animal kingdom has responded to different opportunities for expansion, most notably opportunities opened up by expansion of the food supply provided by plants.

The earliest macroscopic land plants may have resembled modern mosses, liverworts, and hornworts, which have root-like and leaf-like structures but lack a well-developed vascular system to transport fluid between these organs. These plants cannot grow large because they depend on diffusion for fluid transport;

most are only a few inches high. It is not yet known whether these nonvascular land plants were in the main line of descent or represented an evolutionary dead end.

The vascular system of higher plants contributes to the solution of several of the problems of living on land. It is composed of long, rigid cells that are connected end-to-end to provide internal plumbing. This system allows liquids carrying water and nutrients to rise from the roots to the upper levels of the plant. At the same time other parts of the system allow a return flow of liquids rich in the carbohydrate products of photosynthesis. These are the materials that the plant uses for growth and as a source of energy for its metabolic processes. In addition to their role as the circulatory system of the plant, the cells of the vascular system provide strength to the stem. Associated with the vascular system are the roots, which anchor the plant to the ground and absorb moisture and nutrients from the soil. The earliest plants, like algae, liverworts, and mosses, did not have extensive, vascularized, subterranean root systems.

Other important developments associated with the plant vascular system are the cuticle and stomata. These make possible the efficient conservation of water while permitting control over the movement of fluids and dissolved substances within the plant and the exchange of gases with the environment. The cuticle is a waxy coating on the leaves and stems that is impervious to water and prevents the tissue of the plant from drying out. In leaves, the cuticle is penetrated by stomata, small holes that open when the plant needs to exchange gases with its surroundings, particularly to absorb carbon dioxide from the air, but close when the plant needs to conserve its water supply by preventing evaporation from the moist interior of the leaves to the surrounding air. Elements of a vascular system and associated organs were present in some plant species by early Silurian times, and true vascular plants, called tracheophyta, became widespread and abundant during the Devonian period (Figure 12–4). The first vascular plants were probably restricted to moist environments by their method of reproduction. Indeed, much of the later history of land plants is concerned with improvements in methods of reproduction and dispersal.

Organisms with a sexual mode of reproduction usually have two distinctly different kinds of cells. One kind has half as many

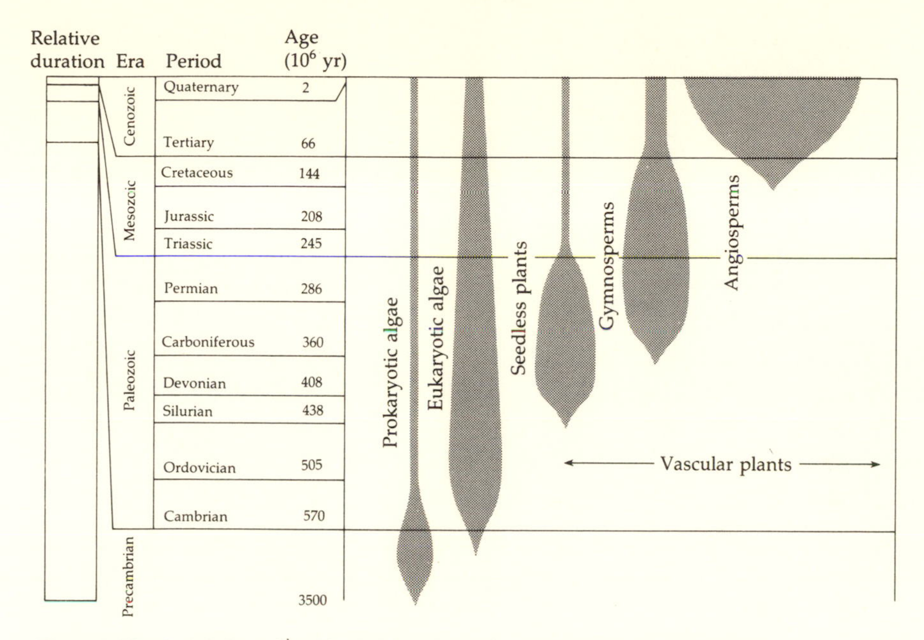

Figure 12–4. Major developments in the evolution of plants.

chromosomes as the other. These are the gametes, cells that join in the process of sexual union. In humans, the sperm and the ova are gametes. In sexual reproduction, two gametes (one from each parent) unite to form a single offspring cell. Each gamete contributes genetic material to the offspring cell, and in this way the offspring cell combines the genes of both parents. Thus the offspring cell necessarily contains twice as much genetic material (twice as many chromosomes) as the gametes, or germ cells, that united to form it (Figure 12–5). In order to start a new generation of cells with the original number of chromosomes it is necessary for the offspring cells to divide, at some stage in the life of the organism, in such a way as to halve the number of chromosomes in the product cells. The term "diploid" is used to refer to cells with the double complement of chromosomes, and the term "haploid" describes cells with a single complement of chromosomes.

The two kinds of cell division that occur during the growth and development of an organism are called mitosis and meiosis. In mitosis, a cell divides in such a way as to produce two offspring cells with chromosomes identical in number and kind to the par-

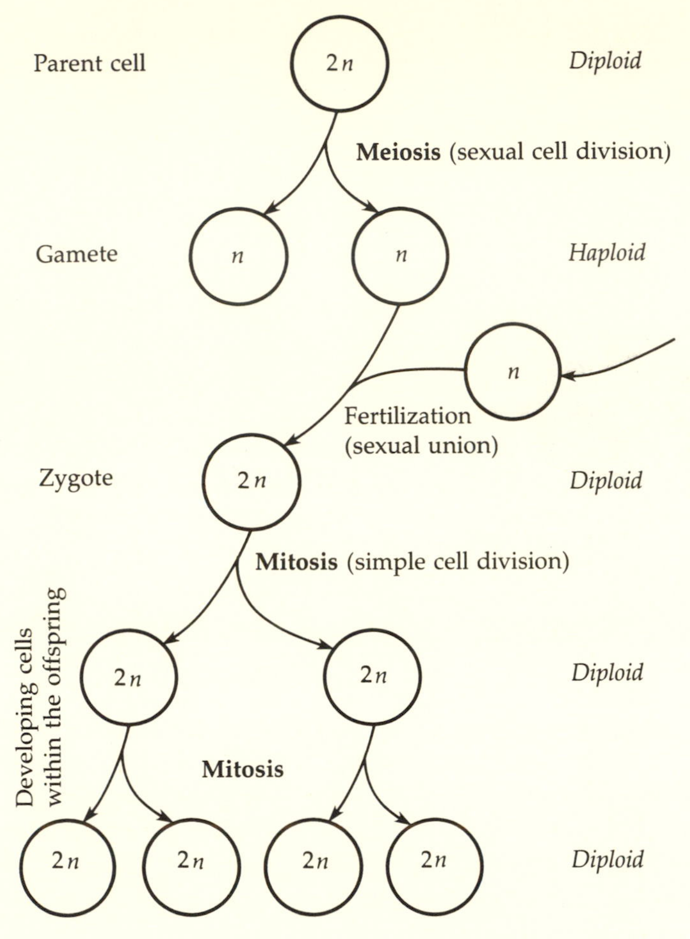

Figure 12–5. Changes in the number of chromosomes in each cell (n or $2n$) during the course of reproduction and development.

ent cell; the number of chromosomes in the cell does not change. In meiosis, the cell, on division, gives half of its chromosomes to one offspring cell and half to the other. Meiosis therefore converts a diploid cell into two haploid cells.

Humans are composed almost entirely of diploid cells. Meiosis occurs only in the production of the reproductive cells, sperm and ova, which are haploid and have 23 chromosomes each. These haploid cells are the gametes mentioned above. The gametes unite when fertilization occurs to produce a diploid zygote that has 46 chromosomes. The cells of the zygote divide by mitosis, without change in the number of chromosomes, in order to produce an embryo and then a child and ultimately another adult human.

Although this arrangement may seem obvious and logical to us, it is not the only possible way to organize a life cycle. In some algae, for example, the free-living organism is haploid. Haploid gametes are produced by mitosis. These join in sexual union to yield a diploid zygote that immediately undergoes meiosis to produce haploid spores. Although haploid, these spores combine genetic information from both of the parental gametes. The spores are dispersed by the wind and under favorable circumstances develop into new haploid individuals. Other algae exhibit a life cycle essentially like that described above for humans. The free-living plant is diploid. Meiosis produces haploid gametes that unite sexually to yield a diploid zygote. The zygote then develops by mitotic cell division into a new organism. Some eukaryotic algae can even reproduce asexually by mitosis, in which a single cell divides to produce genetically identical offspring cells. Most interesting, perhaps, are the plants in which both haploid and diploid stages are free-living individuals. Some algae exhibit this kind of a life cycle, as do seedless vascular plants.

Modern ferns are representative of the seedless vascular plants that played an important role in the adaptation of plants to a life on land. The adult plant is diploid. It produces haploid spores, which are reproductive cells that can survive for long periods of time in a dormant state when conditions are unfavorable for growth. The spores are dispersed by the wind. When they fall on suitably moist ground they develop into a small, haploid plant, called a gametophyte, that is morphologically quite different from the spore-bearing adult (called a sporophyte). The gametophyte develops mitotically from the haploid spore. It is therefore a free-living haploid individual. The gametophyte produces haploid sperm that must swim a short distance to fertilize a haploid egg produced by another gametophyte. This is the stage of sexual union. It produces a diploid zygote, the fertilized egg of the gametophyte, that develops into an adult sporophyte. Because the mobile sperm can move only through a film of water, the seedless vascular plants have not fully adapted to reproduction on dry land. Nevertheless, they were the dominant land plants throughout the Devonian and remained an important component of the terrestrial flora until the end of the Paleozoic. With strong root and vascular systems and abundant leaves they evolved both small herbaceous representatives and large woody trees as much as 25 feet tall. Tree ferns survive to this day in tropical forests.

Seed-bearing plants began to appear in the late Devonian. The basic innovation was the retention of the gametophytic stage of development within the sporophyte rather than as a distinct individual dependent on a moist environment. In seed plants, the mature plant, a diploid sporophyte, produces male and female spores by meiosis, just as do many algae and seedless vascular plants. The spores are not shed, however, but are retained within the moist tissues of the adult plant where they can develop in a benign environment. The female spores develop within a fleshy covering into mature female gametophytes, drawing nourishment from the tissues of the parent sporophyte. The organ consisting of the covering and the enclosed gametophyte is called an ovule. Meanwhile, the male spores develop into male gametophytes enclosed within a tough protective shell. The organ is called a pollen grain. The covering of the pollen grain protects the male gametophyte from hostile environments and desiccation. Pollen grains are dispersed by various means. In the earliest representatives of the seed-bearing plants dispersal was probably by wind or running water.

When a pollen grain lands close to an ovule, the male gametophyte grows a pollen tube that penetrates into the interior of the ovule. It then releases sperm that travel through the pollen tube down into the ovule to fertilize the egg contained within the female gametophyte. The fertilized egg develops into an embryo contained within the tissues of the female gametophyte, which in turn is contained within the ovule. The final product of this process of development is a seed, an embryonic sporophyte surrounded by starchy nutrients. After transport to a favorable environment, these nutrients nourish the embryo during the early stages of its growth into a new plant.

Conifers are representative of the class of plants called gymnosperms or naked seed plants, which also includes seed ferns, cycads, and ginkgoes. In these plants, the seeds develop on the surface or near the tip of an appendage, rather than enclosed within the tissue of a fruit. Pinecones are examples of the seed-bearing structures of naked seed plants.

The development of seed and pollen freed the gymnosperms from a dependence on a moist environment for reproduction. They were therefore able to colonize extensive new areas of land. For reasons that are not understood, however, they seem never to have developed small herbaceous forms, being represented only by large trees and woody shrubs. During the late Paleozoic and early Mesozoic, then, the dominant forest trees were gymnosperms, but the small, nonwoody undergrowth consisted mainly of ferns and other seedless plants.

This situation changed with the rise of flowering plants in the Cretaceous Period. The steps that led to their development are not known. This group is characterized by tremendous structural plasticity and adaptability, which has permitted the development of a wide variety of shapes and habits. The group has evolved large trees, woody shrubs, and small herbs. Some flowering plants, such as cacti, have adapted to extreme aridity while others have returned to an aquatic life. Among the flowering plants there are even parasitic species and insectivores. The factors responsible for the adaptability of flowering plants are not well understood, but they have contributed significantly to the success of this group. Out of a total of 260,000 living species of vascular plants, 250,000 are flowering plants, only 700 are gymnosperms, and most of the remainder are ferns.

Improved methods of reproduction also contributed to the success of the flowering plants, of course. The organs that produce seed and pollen are incorporated into flowers that attract animal and insect pollinators. This approach to cross-fertilization is more reliable and less wasteful than reliance on the wind because the pollinators are likely to travel directly from one plant to another of the same species. At the same time, the seed is enclosed within the structure of the flower and can develop a covering that promotes dispersal. Examples of such coverings include fruits, prickly burs, and the fluffy sail of the dandelion seed. Flowering plants are formally called angiosperms in reference to their covered seeds.

On the animal side, it appears that our closest ancestors in the late Precambrian were flatworm-like creatures, possibly descended from planktonic larval jellyfish. These creatures ate organic detritus and lived at the bottom of shallow seas. A number of architectural developments enabled them to prosper. First they acquired a body cavity, called a coelom, that could be filled with fluid to provide some rigidity to the structure of the organism. By allowing forces generated by the muscles of the entire body to be concentrated on a small area, this hydrostatic skeleton permitted these creatures to burrow after food in soft sea floors. Traces of their burrows have been found in 700-million-year-old sediments.

The locomotory power provided by the coelom opened up new ways of exploiting the food resources of the sea floor. There arose active burrowers that exploited buried food, sessile (stationary) suspension feeders that captured the food before it settled to the bottom, and creepers that grazed on the surface of the mud. Adaptation to these markedly different ways of life led to markedly different morphological developments. All of the subsequent elaboration of animal life is based on one or another of the major bodily architectures established in the shallow seas of the very late Precambrian.

The appearance of mineralized skeletons early in the Cambrian period represented a distinct improvement in locomotory capability. Skeletons could be articulated and the different portions connected by muscles. The first skeletons were on the outside of the animal and so were potentially useful also as a defense against predators. There is a classical debate, not yet resolved, over whether special environmental circumstances are required to explain the apparently sudden appearance of mineralized skeletons in the fossil record.

On one side it has been argued that there must have been a long history of evolution of soft-bodied and rarely fossilized metazoa in the later Precambrian about which we know little. Some change in the chemical or physical environment at the beginning of the Phanerozoic made it possible for the first time for organisms to secrete shells. The argument on the other side is that metazoa originated quite late in the Precambrian and expanded rapidly into vacant ecological niches. The proliferation of skeletons was one consequence of this adaptive radiation, but it was also an important cause of it: Mineralized skeletons made new ways of life possible for animals.

Recent, careful examination of the fossil record of the first metazoa appears to support the latter interpretation. A number of important groups of metazoa originated without skeletons during the last 50 million or so years of the Precambrian. During an approximately equal period of time at the beginning of the Cambrian these groups acquired skeletons. The appearance of skeletons was no more simultaneous than the appearance of the major groups of metazoans themselves. It seems likely that the timing was determined not by environmental change but simply by the pace of biological evolution.

The next major advance was the origin of the backbone (Figure 12–6). The first vertebrates were fishes, which appeared near the beginning of the Ordovician Period. A light and flexible interior skeleton permitted greater mobility than a bulky exterior skeleton. The development of the jaw during the Silurian period gave fishes a stronger bite than that of any preexisting organisms. One group of fishes developed stubby, muscular fins on which to walk along the bottom of the sea. During the Devonian, this group acquired a respiratory apparatus that enabled its members to survive out of the water for short periods of time. These fishes took to making excursions onto the land either to look for food or to escape from predators. Remember that plants were the first organisms to make the transition to land, becoming established there by the early Silurian.

Animals were quick to follow the plants ashore. Insects, spiders, and snails appeared on land at essentially the same time as plants. They presumably went there to eat the plants and then to eat one another. By Carboniferous time there was a diverse insect fauna including dragonflies with a wingspan of 30 inches

NICHES AND OTHER ECOLOGICAL CONCEPTS

In ecology, a population *is a group of individuals of the same species, generally inhabiting a specific region. The physical living space of a population is called its* habitat. *The habitat is the "address" of the population; it says where it is to be found. The habitat of professors, for example, is universities.*

The niche, *in contrast, describes the way the population goes about the business of living, how the individuals in the population get food, shelter, offspring, and whatever else they need to survive and reproduce. The niche is the "profession" of the population. The niche of professors, narrowly defined, is teaching and research.*

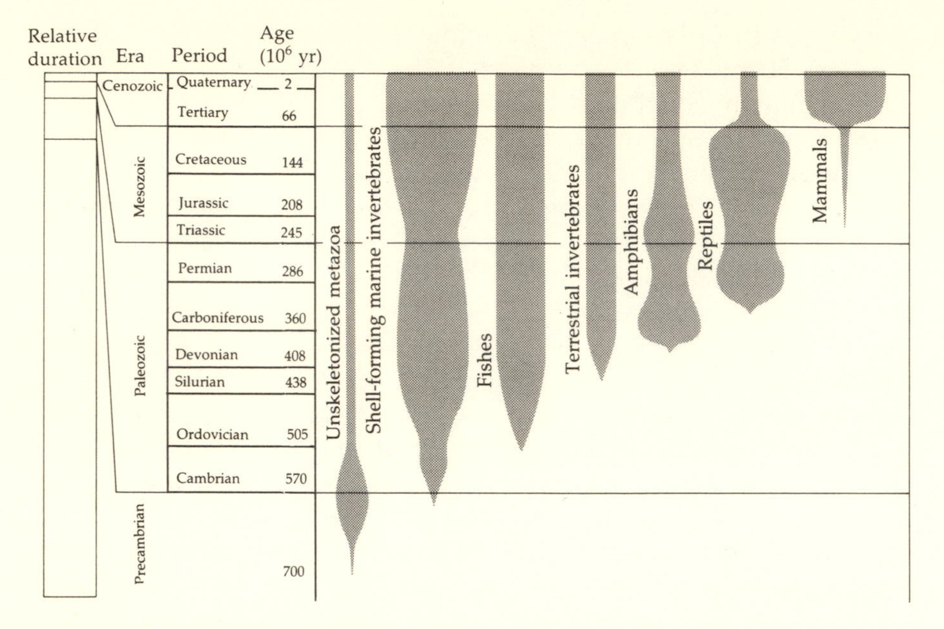

Figure 12–6. Major developments in the evolution of animals.

and cockroaches 4 inches long. The arrival on land of our vertebrate ancestors was delayed, however, until the late Devonian, when the first amphibian (called labyrinthodont because of the labyrinthine infolding of its tooth enamel) made its appearance.

Apart from breathing, the main problem that large animals had to overcome in the transition to land was that of structural support. The skeleton had to support the weight that had been supported by buoyancy in the aquatic environment. Much of the evolution of terrestrial vertebrates has involved the reorganization of the skeleton to provide more support without loss of mobility. The earliest amphibians, which appeared in the late Devonian, looked very much like fishes, but their bodies and heads quickly became flattened and their limbs grew shorter. Their eyes moved to the tops of their heads, suggesting that they spent much of their time in shallow water.

Amphibians were tied to water because they produced eggs like those of fishes, in which the embryo is encased in a thin membrane that allows oxygen and waste matter to be exchanged with the surrounding water. Such eggs dry out if exposed to air. Reptiles solved this problem in the Carboniferous period by developing a large egg with a hard covering that protects the contents from desiccation, a supply of food for the embryo (the yolk), and a storage area for waste material. This adaptation freed reptiles to range over a much larger area than the amphibians, and in time reptiles became the dominant group.

The Mesozoic era was the well-known age of reptiles. Reptilian forms occupied a large number of ecological niches on land, and some even became adapted to flight while others returned to the sea and took up the way of life of carnivorous fishes. The dinosaurs are the most familiar of the Mesozoic reptiles. They arose in Triassic time from a small, lizard-like reptile called a thecodont, flourished during the Jurassic and Cretaceous periods to the point where more than 230 genera are known, and disappeared, suddenly and mysteriously, at the end of the Cretaceous.

Reptiles were succeeded as the dominant land animals by mammals. Mammals arose during the Triassic period from a reptilian group known as therapsids. The first forms were small creatures resembling rats, and mammals remained small and inconspicuous throughout the Mesozoic. Evidently their more careful nurturing of the young did not give them any significant advantage over competing reptiles. Some aspect of mammalian adaptation did, however, bring them unscathed through the biological crisis that wiped out the dinosaurs and many other reptilian lines at the end of the Mesozoic era. In the Cenozoic, mammals assumed the ecological roles previously played by reptiles so completely that some forms once again took to the air (bats), while others returned to the sea (whales, dolphins, and porpoises). It seems fairly clear that the mammalian expansion had to wait for the opening up of ecological niches by the extinction of the dinosaurs. What factors of change in the global environment, evolution of ecosystems, or extraterrestrial catastrophe were responsible for the decline of the reptiles remains subject to hot debate.

The replacement of dinosaurs by mammals provides a clear illustration of a very important concept. Evolution is opportunistic. Organisms have the genetic resources to exploit new opportunities rapidly when they appear, but a biological innovation at the wrong time goes nowhere. Mammals flourished only after the dinosaurs had disappeared. There is no evidence that the dinosaurs were ecologically inferior to the mammals, except for their failure to survive the biological crisis at the end of the Mesozoic. Without that crisis, the mammals might never have had their day in the sun. In the same way, any fish that grew legs before there were plants on land would have had nowhere to go. Flowers and fruit were no use to plants before there were animals and insects on land to be attracted by them.

Many flowering plants are remarkably adapted to their pollinating insects or animals. Flowers are often configured in such a way that only a single species of pollinating organism can reach the nectar and pick up pollen. In the same way, there are insects that are completely dependent on a single species of flowering plant. This remarkably specialized adaptation of flowering plants with their pollinators is a consequence of coevolution. Over a long period of time evolutionary changes in the flowers have stimulated corresponding changes in the pollinators and the other way around. Over a period of time, in this way, plant and pollinator have become more finely adapted to one another.

More than anything else, it is probably the opportunistic quality of evolution that determines the pace as well as the course of biological change. The transition from fish to amphibian, for example, was not made in one jump. The organisms that took part in this transition were led along an adaptive pathway by a particular sequence of ecological opportunities (Figure 12–7). If this sequence had been different the end product would probably have been different too. Alternatively, the mechanism of inheritance might have come up with different answers to some of the ecological problems posed by the environment. The course of evolution has been guided by an exceedingly subtle and complex interplay of environmental determinism and genetic chance, an interplay that determines strategies for exploitation of the natural economy. It is not at all clear that the outcome would be even approximately the same if the experiment were to be run again. The pace of biological change is probably governed more by the appearance of exploitable opportunities than by the rate of mutation.

Nevertheless, it is clear that the over-all direction of evolution is such as to use more fully and efficiently the resources of the biosphere. My account of the history of life as revealed by the fossil record shows this. Since the beginning of the Phanerozoic, life has expanded into previously unoccupied terrain and has

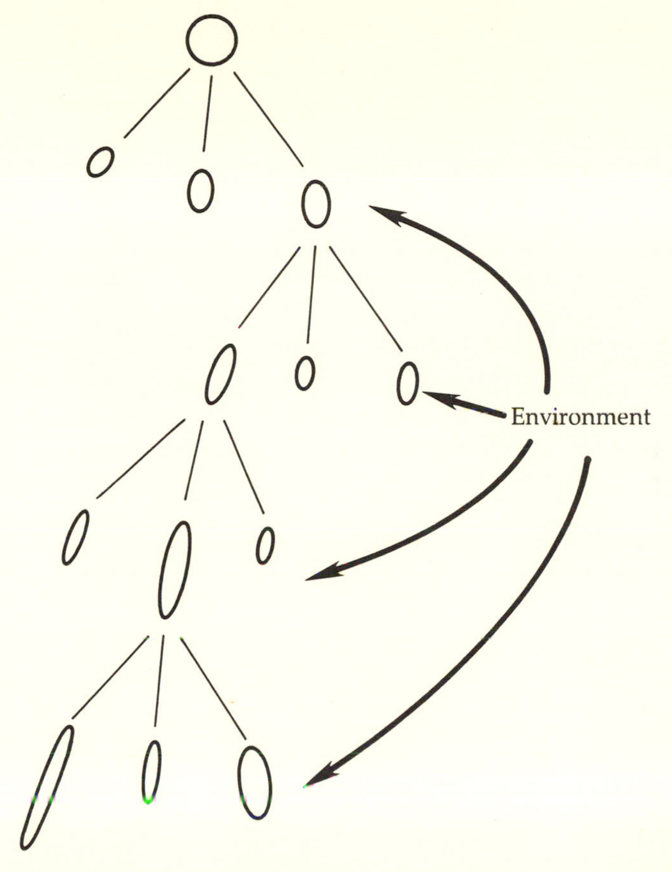

Figure 12–7. The direction of evolution is controlled by the interaction of the environment, including other species, with differences among individuals in successive generations.

evolved communities of organisms of increasing complexity to make the fullest possible use, consistent with ecological constraints, of the resources of each of its ecosystems. This trend is entirely consistent with our understanding of the mechanism of evolution. Any habitat or resource going to waste, any ecological niche unoccupied, offers less competition and therefore more reproductive success to the species that can move in.

So far I have provided general answers to the questions posed earlier: what have been the key developments in the history of life; what is the mechanism of change; what determines the direction and the pace of evolution? But in one important respect the story developed so far does not correspond to the evidence of the fossil record. The traditional view of evolution is of a process proceeding gradually and steadily as natural selection modifies morphology to keep organisms adaptively tuned to their environments. The picture that emerges from the fossil record is rather different. Although there are exceptions, most species show little change in morphology during their existence. They do not evolve gradually into new species. Instead they appear suddenly and disappear equally suddenly, to be replaced by new species (Figure 12–8). The changes involved in the origin of higher taxonomic categories, genera, families, orders, classes, and phyla, are even more marked. Are these apparent discontinuities a consequence of gaps in the historical record or the immigration of new species that have evolved elsewhere, or is evolution truly episodic, possibly reflecting an episodic compo-

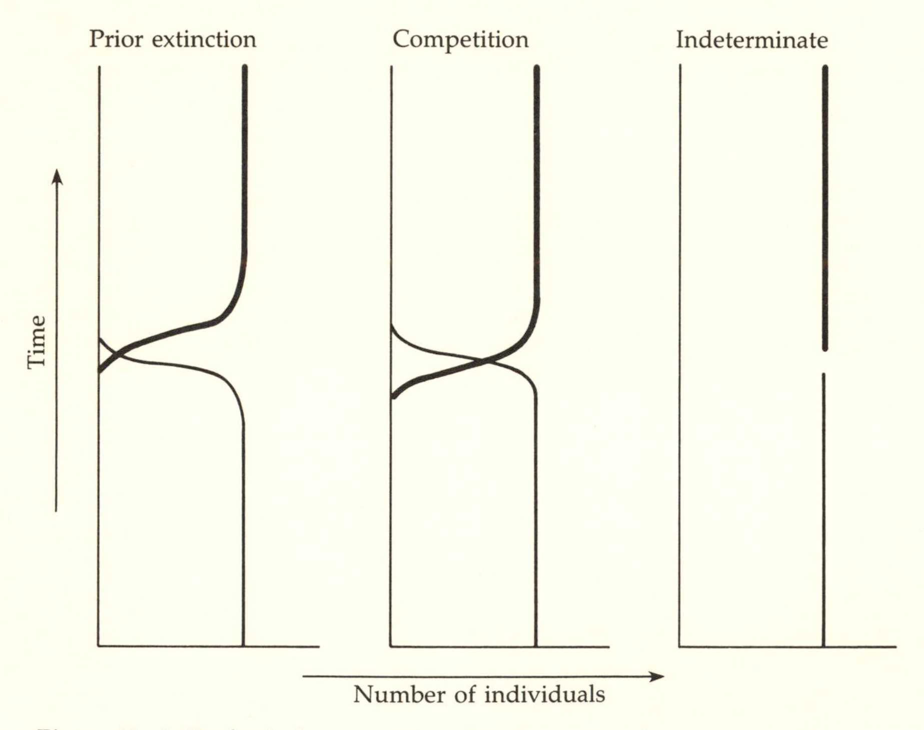

Figure 12–8. Ecological succession. Good time resolution is needed if the geological record is to reveal the mechanisms of change in the species occupying a particular ecological niche.

nent of change in the physical and chemical environment of life? This question is still being debated, but most paleontologists probably feel that the known fossil record is a reasonably accurate record of the pace of biological change.

Long periods of relative quiescence have been punctuated by short episodes of rapid biotic change. The beginning of the Cambrian, for example, was a time of pronounced biological experimentation, when many new species arose. The ends of both the Paleozoic and Mesozoic eras were marked by biological crises in which many species and higher taxonomic categories, including the dinosaurs, became extinct, and many new species appeared. To a lesser extent, the end of each geological epoch was characterized by biological change, frequently at an accelerated rate. The cause of this uneven pace of evolution is a popular subject of speculation, particularly insofar as it touches on the fate of the dinosaurs. To some extent, of course, every markedly new biological adaptation should lead to a period of rapid biological change as new species evolve with the new adaptation, and old species become extinct as a result of competition from the new species. But this mechanism does not explain the major crises in the history of life, such as those at the end of the Paleozoic and Mesozoic eras, when new species appeared in the fossil record only after the disappearance of the old species.

Mass extinctions are a reality in the history of life. They have probably been caused by changing environmental conditions brought about, in many instances, by the changing distribution of the continents, a result of continental drift. The message of evolution is that some lineages can adapt to changing conditions; they survive. Other lineages are not able to adapt fast enough and become extinct.

Suggested Reading

Bold, H. C. *The Plant Kingdom.* Englewood Cliffs, NJ: Prentice-Hall, Inc., 1977.

McAlester, A. L. *The History of Life.* Englewood Cliffs, NJ: Prentice-Hall, Inc., 1968.

Scientific American. *Evolution and the Fossil Record.* San Francisco: W. H. Freeman and Co., 1978.

CHAPTER 13

Good Times And Bad

THE HISTORY of life revealed in the fossil record is a history of constant change. Why has the biota as a whole never become so well adapted to its environment as to make further change unprofitable? In the discussion of Precambrian evolution in Chapter 6, I attributed metabolic change to inadequacies in the supply of organic molecules that were synthesized by nonbiological processes in the ocean and atmosphere. The organism that developed the ability to synthesize whatever foodstuff was in short supply gained a competitive advantage over other members of its community. Competition has undoubtedly been a major source of evolutionary change during the Phanerozoic as well. The first amphibians—fishes that acquired the ability to survive and move on land for limited periods of time—found a source of food and a refuge from predators that was denied to their wholly aquatic relatives.

Changes in the structure and organization of the biosphere have provided a strong stimulus to biological evolution as well. There have been developments, some biological and some not, that have at times made it possible for more species to inhabit the globe. The biota has responded to these expanded opportunities with waves of biological innovation. For example, an adaptive radiation and expansion of life followed the origin of metazoa late in the Precambrian. At other times the carrying capacity of the biosphere has declined, not necessarily in terms of numbers of individuals but in terms of numbers of species. These declines have frequently been associated with geographic changes such as those to be described in this chapter. The consequence of a decline in carrying capacity has been a decline in the diversity of life (the total number of species), which necessarily implies the extinction of some species. Subsequent amelioration of whatever factors caused the decline in carrying capacity has permitted diversity to increase again, with the appearance of new species and opportunities for radical biological innovations. For example, a large increase in the number of species of mammals followed the mass extinction of reptiles at the end of the Mesozoic, about 65 million years ago.

The impact of changing conditions can be seen most readily in the fossil record of the well-skeletonized invertebrate species that lived in shallow water on the continental shelves. By counting the number of taxa of these organisms identified in rocks of different ages it is possible to learn how their diversity has varied. The record shows that diversity was low at the beginning of the

TAXONOMY

Organisms are classified on the basis of their resemblance to one another into categories in a system first proposed by the eighteenth-century Swedish botanist, Carolus Linnaeus (1707–1778). The lowest order taxonomic category is called a species. It consists of a group of individual organisms sufficiently similar in genetic makeup to be able to interbreed. All domestic dogs, for example, constitute a species called Canis familiaris. *The first word in the Latin name identifies a grouping of similar species called a genus (plural, genera), and the second word distinguishes the species. Another member of the genus* Canis *is the coyote,* Canis latrans.

Just as similar species are grouped into genera, so similar genera are combined into families, families into orders, orders into classes, and classes into phyla. Modern humans, for example, are the species sapiens *of the genus* Homo *of the family* Hominidae *of the order* Primates *of the class* Mammalia *of the phylum* Chordata. *A particular level within this hierarchical structure, for example families, is called a taxonomic category, and a group of organisms classified as a unit within a category forms a taxon (plural, taxa). The* Hominidae, *for example, are a taxon.*

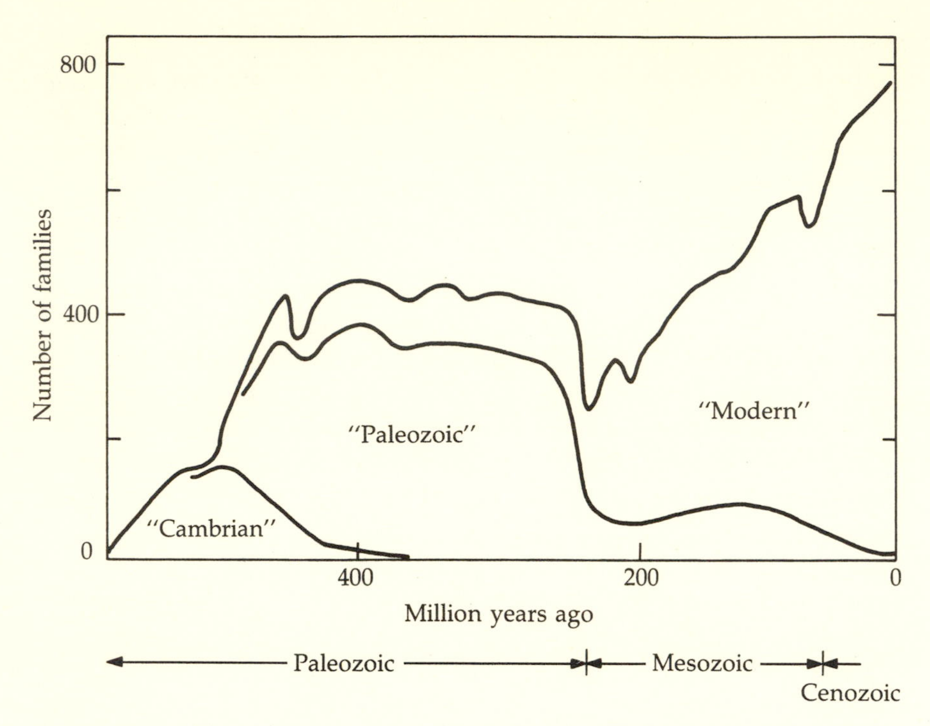

Figure 13–1. Fluctuations in the diversity of well-skeletonized shallow-water marine invertebrates through the Phanerozoic. (After J. J. Sepkoski, *Paleobiology, 7,* 36, 1981.)

Paleozoic, as is to be expected because these organisms had only just originated. The diversity increased rapidly as new species evolved, and achieved a plateau in the Ordovician (Figure 13–1). Fluctuations during the rest of the Paleozoic were minor. Some taxa became extinct, but they were replaced by approximately equal numbers of new taxa. At the end of the Permian, however, there was a marked drop in diversity, indicating that many lines became extinct at about the same time and that replacements took some time to appear. Diversity recovered during the Mesozoic to levels considerably higher than those of the Paleozoic and has remained high ever since.

Why is there a limit to the number of species that inhabit the earth or any region of the earth? Why does the biosphere have a carrying capacity? The answers derive from the concept of ecological niche. Two species cannot for long live in the same place making their livings in identical ways and competing for identical resources. In such a situation one species expands in population at the expense of the other, depriving the subordinate species of resources and driving it to extinction. In order for two species to coexist in the same ecosystem there must be some differences in their utilization of resources. For example, they must feed at different times or prefer different foods. Each species needs its own ecological niche.

The number of niches in a given area depends partly on biological factors and partly on environmental factors. For example, a given area of seashore offers more heterogeneous opportunities for different ways of life than an equal area of grassland. As an example of the role of biological factors, consider the community of animals that eat detritus on level mudflats under water. Creeping organisms can collect their food only at the interface between mud and water. Increased diversity is possible if the community includes burrowing organisms than can harvest buried detritus missed by the surface feeders. A further increase in diversity could result from the introduction of taller suspension feeders that capture detritus before it settles to the bottom. To some extent, then, new organisms create new ecological niches, and one aspect of the history of life has been the increasingly fine subdivision of ecosystems into niches occupied by distinct species. Empirical observation indicates, however, that there are practical limits to this process that are not clearly understood.

Ecological studies have revealed a number of other, nonbiological factors that limit species diversity. One of these is environmental heterogeneity. The diversity of plants in a given area of level grassland is lower than that of an equal area of hilly country, where conditions of climate, nutrient supply, and soil moisture vary with altitude. Another factor is the area of the ecosystem, an effect illustrated most clearly by studies of diversity on islands (Figure 13–2). Small islands support fewer species per unit of area than large islands. This effect may be related to population sizes. A small area can support only a relatively small population of a given species. A small population is less likely to survive a period of stress (drought for example) than a large one. A possibly related factor involves the abundance of resources. Diversity may be larger where food (for example) is abundant than where food is in short supply, because abundant food permits large populations and large populations are less likely to die out.

A particularly important consideration is the stability of the

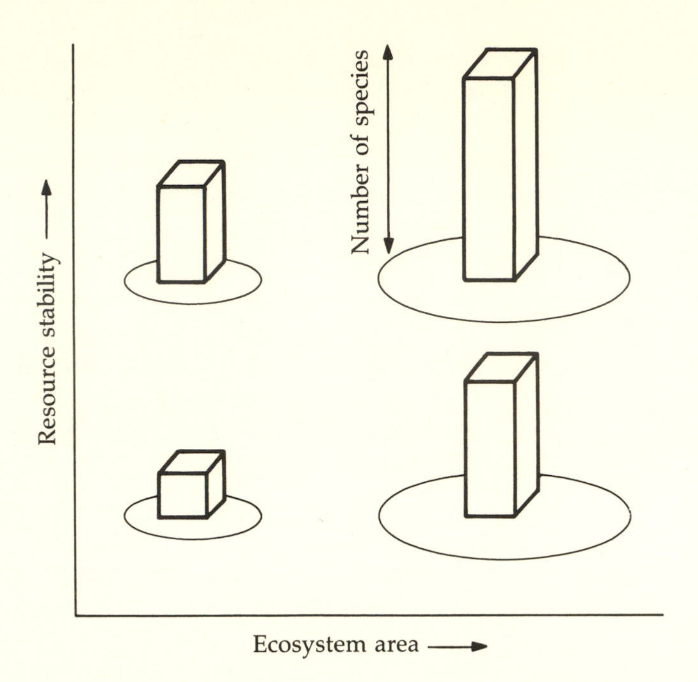

Figure 13–2. Diversity increases with increasing stability of resources and increasing area of the ecosystem.

environment. Its impact on diversity is clearly revealed by studies of the animals that live on the continental shelves today. The most diverse shallow water faunas are in the tropics, where communities consist of large numbers of highly specialized species. Diversity gradually decreases with increasing latitude. In the arctic zones there are fewer than one-tenth as many species as in the tropics, when comparable areas are considered. The diversity gradient correlates well with a gradient in the stability of food supply. Seasonal variations are small in the tropics, where food is produced by oceanic autotrophs (phytoplankton and algae) at a relatively constant rate. This stability enables some organisms to become specialized and to develop narrow preferences for food and habitat as well as specific environmental requirements for wave action and illumination, for example. A stable supply of food also permits the development of a long food chain with many levels of larger organisms preying on smaller organisms all the way down to the primary producers.

As the seasons become more pronounced at higher latitudes, fluctuations in primary productivity—the food supply—become greater. Food chains must be short when resources are fluctuating because it takes time for any expansion in resources to work its way up the chain, and the organisms on the higher levels of a long chain do not have time to benefit before a decline sets in. At the same time, fluctuating resources favor generalized organisms able to exploit different resources at different times of year and tolerant of changes in conditions.

Although the distribution of diversity in today's world is dominated by this latitudinal variation, there are significant differences between areas at comparable latitudes as well. These differences also correlate with the stability of the food supply. Diversity is lower along coasts where there are marked seasonal changes in the currents or in the upwelling of deep water, for example, that affect primary productivity by changing the supply of nutrients to plankton. At any given latitude, therefore, diversity is highest off the shores of small islands in large oceans where seasonal changes are small, and least off the shores of large continents facing small oceans where the continental landmass has a destabilizing effect on the climate. The modern oceans therefore provide evidence for the dependence of diversity on environmental stability.

These factors—heterogeneity, area, resource supply, and stability—influence the diversity of a given region. For the total diversity of the global biota we must consider also the number of regions, called provinces, into which the biosphere should be divided. This consideration is environmental heterogeneity on a global scale. As an illustration, consider again the shallow water fauna of the modern oceans. They are highly provincial, which means that species living in different oceans or on different sides of the same ocean tend to be quite different. The reason for this, of course, is that most shallow marine organisms cannot disperse over land or over large distances of open ocean. Climate also causes major variations in species composition, even along continuous coastlines. The distribution of the continents today has produced an unusually large number of different provinces. For one thing, the north–south orientation of many of the continental coastlines means that each continent can support a full range of climatic provinces along each coast. At the same time, the isolation of the polar regions from the rest of the world ocean has caused an unusually large variation of temperature with latitude, so the number of different provinces that can exist along a given north–south coastline is high. These factors combine to divide the shallow water biota of today into 30 or so distinct provinces, all of which share a relatively small proportion of their species. The resultant total diversity is more than 10 times as high as the diversity of a single province, even a very diverse one.

The world at the close of the Paleozoic era (225 million years ago) was very different. At that time nearly all of the land masses were joined together in a single supercontinent called Pangaea (Figure 13–3). There were few physical barriers like open ocean to prevent the dispersal of shallow water marine organisms along the entire continental shelf. Provinciality was therefore low and entirely due to climatic effects. The free motion of ocean currents between the equator and the poles should have led to a reduced latitudinal temperature gradient on the Permian world, so even climatic provinciality was less developed than it is today. These two factors alone would have permitted much less diversity at the end of the Permian, which is what the record shows (Figure 13–1).

A large continent like Pangaea should produce a variable climate in the near-shore areas, leading to instability in the food supply of the shallow marine fauna. This instability depressed diversity even further. As a result, the wave of extinctions that marked the end of the Paleozoic era was the most severe ever suffered by marine fauna. Late Paleozoic populations included many elaborately adapted species that seem ecologically similar to species now found in stable tropical environments. The species that survived into the Mesozoic were much simpler in form and appear to have been bottom-dwelling scavengers or suspension feeders. They were ecologically similar to the populations found today in unstable environments at higher latitudes. As James W. Valentine describes the change, "Late Paleozoic faunas were showy, whereas early Triassic faunas were grubby" (Valentine 1973, p. 456).

Pangaea broke up during the course of the Mesozoic and Cenozoic as the continents moved into their present configuration. By the end of the Triassic period, 180 million years ago, the northern supercontinent (Laurasia) had separated from the southern supercontinent (Gondwana) and Gondwana had already broken into three fragments, Africa–South America, Antarctica–Australia, and India. The Indian subcontinent was moving rapidly northward toward Laurasia. By the end of the Jurassic period, 135 million years ago, both North and South Atlantic Oceans had begun to open from the south, although the northern ends of North America and Europe and of South America and Africa were still connected. By the end of the Cretaceous period, 65 million years ago, the South Atlantic had completely isolated South America. The North Atlantic was wider, but North America was still connected to Europe by way of Greenland. The African plate was impinging on the European plate to produce the Alps. The Indian plate began the collision with Asia that produced the Himalayas about 45 million years ago, and by 35 million years ago Australia had separated from Antarctica and was drifting northwards to its present location. Meanwhile, the North Atlantic broke through to the Arctic Ocean to the east of Greenland, separating North America from Europe, and rotation of North and South American plates brought these two continents closer together. Only about 2 million years ago did North and South America become connected by a land bridge formed by the growth of a chain of volcanic islands, a northward extension of the Andes mountains produced by subduction of the Pacific plate.

This breakup of Pangaea produced smaller landmasses, and smaller landmasses resulted in more stable environments, so communities at low latitudes could be filled with numerous specialized populations. At the same time, new provinces appeared as continents became separated by ocean and as latitudinal temperature gradients increased. In this way the diversity of the late Paleozoic was recovered and even surpassed.

There were also profound changes in geography during the Paleozoic era (Figure 13–4). The era opened with at least six continents located at equatorial latitudes. These continents were quite different from the modern continents formed by the breakup of Pangaea. The largest was Gondwana, composed of land masses that were to become Africa, South America, Australia, Antarctica, and portions of Europe and Asia. High latitude regions were largely free of land, so there were two open polar oceans. Gondwana drifted across the South Pole during the course of the

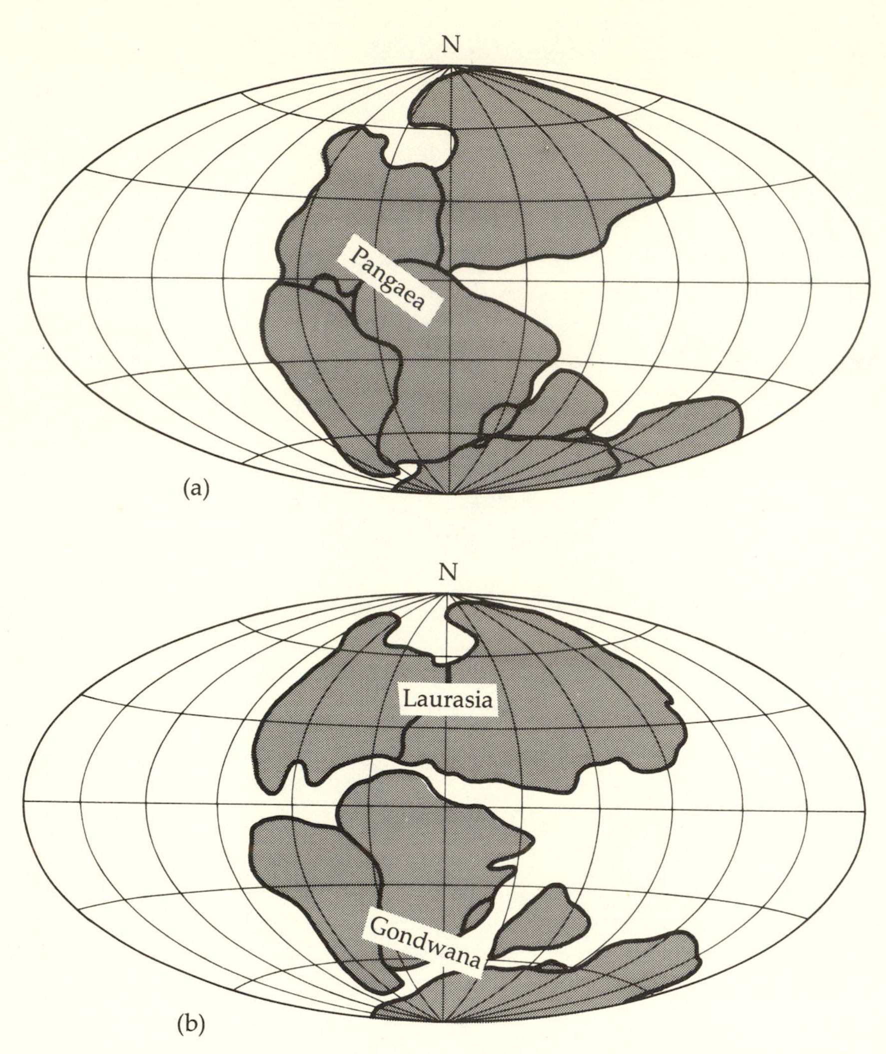

Figure 13–3. The breakup of Pangaea. (a) 200 million years ago; (b) 180 million years ago; (c) 135 million years ago; (d) 65 million years ago.

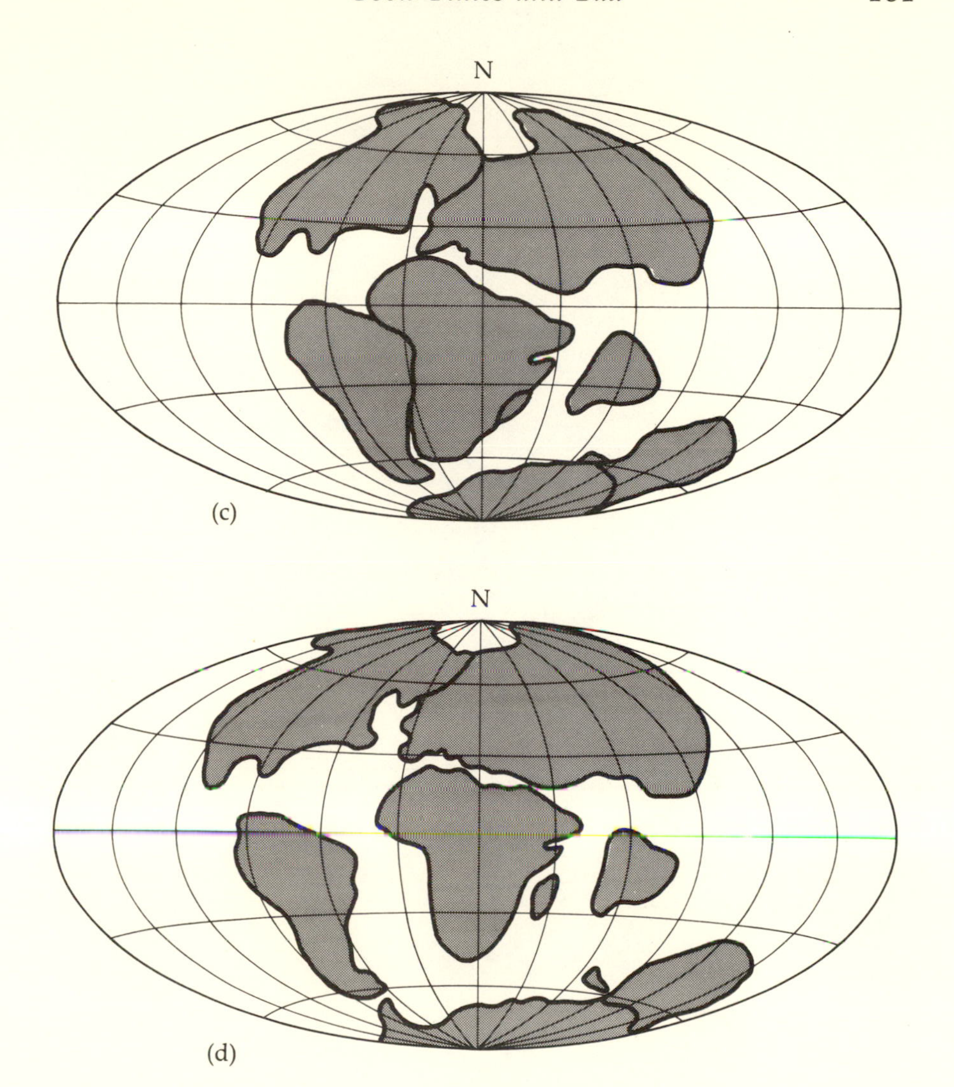
N
(c)
N
(d)

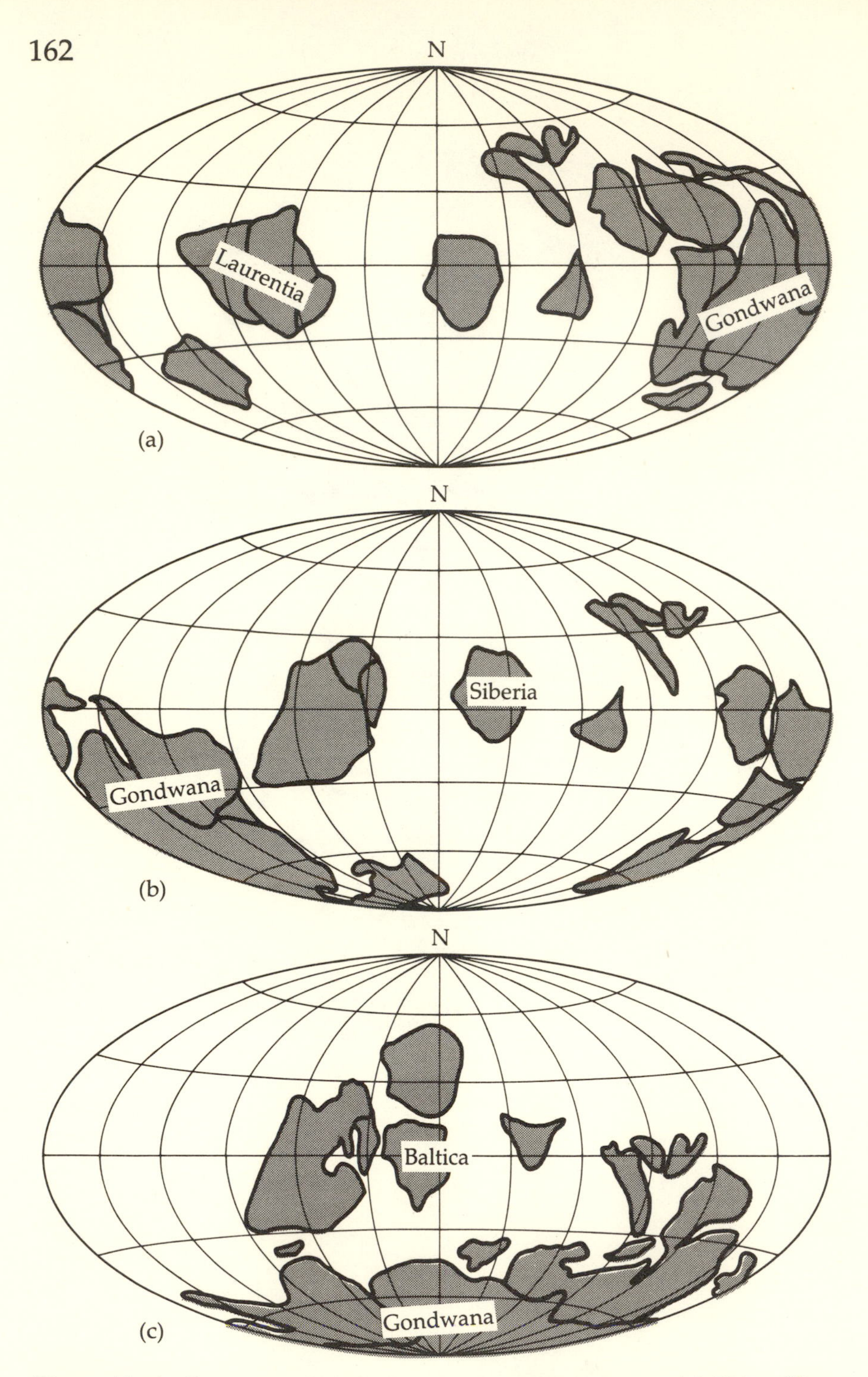

Figure 13–4. Continental drift during the Paleozoic era. (a) 500 million years ago; (b) 450 million years ago; (c) 400 million years ago; (d) 350 million years ago; (e) 300 million years ago; (f) 250 million years ago.

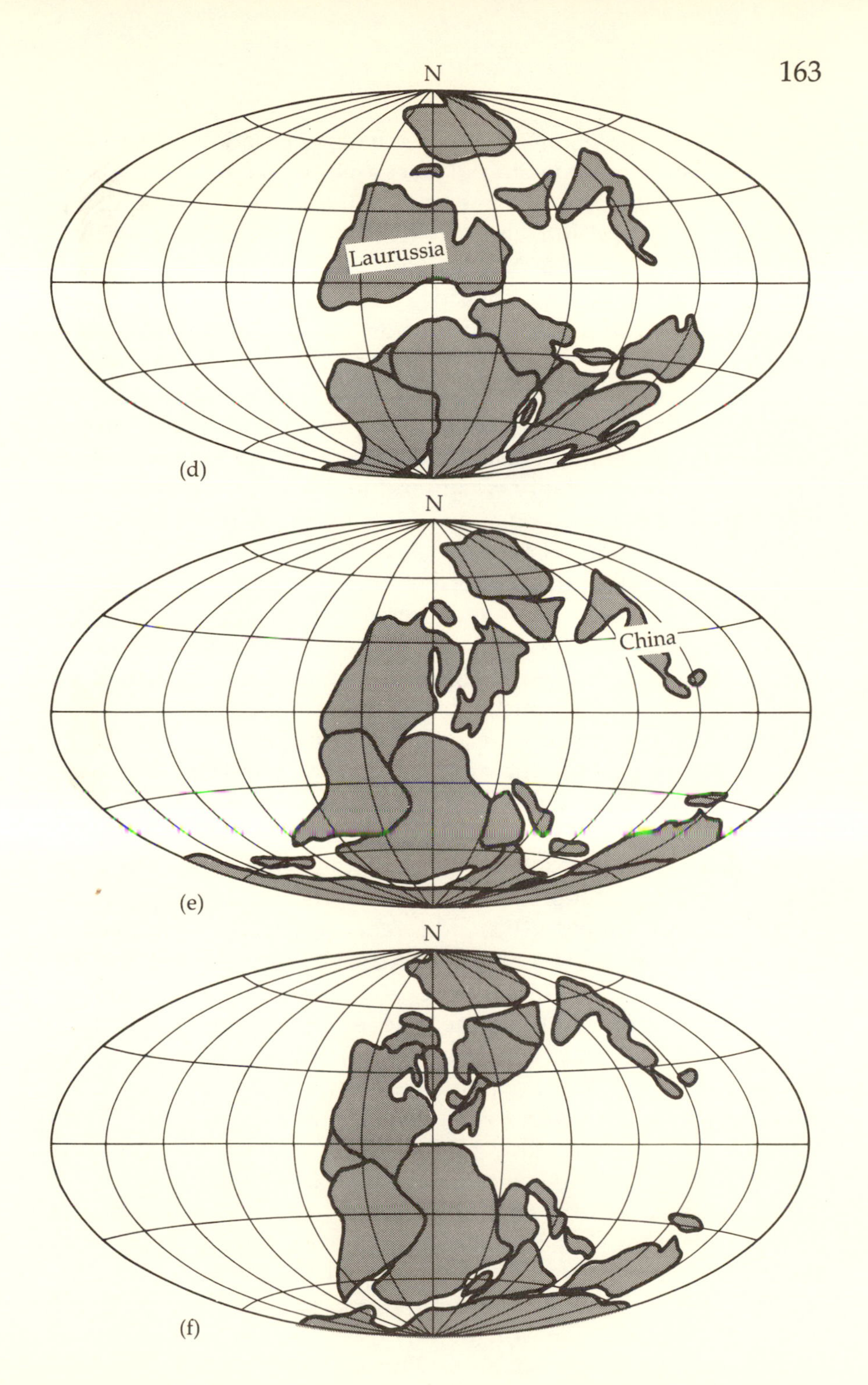
N
Laurussia
(d)
N
China
(e)
N
(f)

Paleozoic and was subjected to recurrent glaciations, but open ocean persisted in the north polar region until late in the Permian (250 million years ago). By the end of the Paleozoic most of the land was assembled into the supercontinent of Pangaea, stretching all the way from the North Pole to the South Pole. An enormous open ocean covered the side of the globe opposite Pangaea. It reached polar latitudes in both hemispheres and covered 300 degrees of longitude along the equator.

One further topic of interest is the effect of the breakup of Pangaea on the terrestrial vertebrates. The reptiles rose to prominence early in the Mesozoic era, when there were few isolated land masses. Provinciality was therefore low. During the 75 million years of the Cretaceous period, for example, there were only a dozen or so orders of reptiles, and most of these orders were represented on all of the modern continents. At the beginning of the age of mammals, however, there were eight separate continents, and evolution followed somewhat different paths on each one. As a result, perhaps 30 different mammalian orders evolved during the 65-million-year history of the Cenozoic, and many of these orders were present on only one continent.

The mammalian fauna of Africa provide an example of what happened next. Until about 25 million years ago, Africa was separated from the rest of the world and many of its animals existed only there. Northward movement of the Africa–Arabian plate established a connection between Africa and Eurasia early in the Miocene epoch (about 20 million years ago), and a number of Eurasian animal species entered Africa. Their competition resulted in the reduction and even extinction of some of the indigenous African fauna. At the same time, ancestral elephants, which had evolved in Africa, crossed into Eurasia and spread rapidly around the world. The elephants had to wait until the end of the Pliocene, about 2 million years ago, before crossing into South America. At that time a land bridge, the Isthmus of Panama, was established by Andean volcanoes and a wholesale exchange of fauna between North and South America took place. South America had previously been almost completely isolated for about 65 million years and had developed a highly unique fauna. Many of the animals were primitive marsupials that could not survive the competition provided by more highly evolved placental mammals of North America. The marsupials became extinct after the land bridge was established. Prior to the junction there were about 29 families of mammals living in South America and about 27 quite different families in North America. The union of the two continents left them with 22 families of mammals in common.

Amalgamation of the continents, migration, and competition have, of course, resulted in a loss of diversity. Approximately 30 orders of mammals evolved in the eight provinces of the early Cenozoic. Today there are only four provinces of land mammals, and 13 orders have become extinct. Although the fragmentation of the continents at the beginning of the age of mammals promoted variety, the subsequent amalgamation has promoted survival of the fittest.

Suggested Reading

McElhinny, M. W., and D. A. Valencio, Eds. *Paleoreconstruction of the Continents.* Washington, DC: American Geophysical Union, 1981.

Skinner, B. J., Ed. *Paleontology and Paleoenvironments.* Los Altos, CA: William Kaufmann, Inc., 1982.

Valentine, J. W. *Evolutionary Paleoecology of the Marine Biosphere.* Englewood Cliffs, NJ: Prentice-Hall, Inc., 1973.

CHAPTER 14

Species Are Not Immortal

THE THEORY of plate tectonics and continental drift provides a powerful tool for understanding the broad patterns of biological change. The impact of continental dispersal and aggregation on the diversity of life and even on the nature of the organisms that were successful at a particular time in the past can be read in the combined geological and fossil records. Indeed, paleobiology and geology are unified by this theory as they have never been before. Factors other than continental configuration have also affected the diversity of life. What does the fossil record reveal about the influence of biological organization, area, and resource supply, for example? Changes in diversity result, of course, from imbalances between the rate of origin of new species (speciation) and the rate at which old species become extinct (extinction). On theoretical grounds we might expect that these rates depend on diversity (Figure 14–1). First, because new species originate from old species we might expect the total rate of introduction of new species to increase with diversity. In contrast, the probability that an existing species will spawn a new species in a given period of time should decrease with increasing diversity because of reduced opportunities for the invasion of unoccupied ecological niches. At the same time, the probability that a species will become extinct during a given time should increase with increasing diversity because of increased competition and the smaller populations that must result from an increased number of species occupying a world of finite resources.

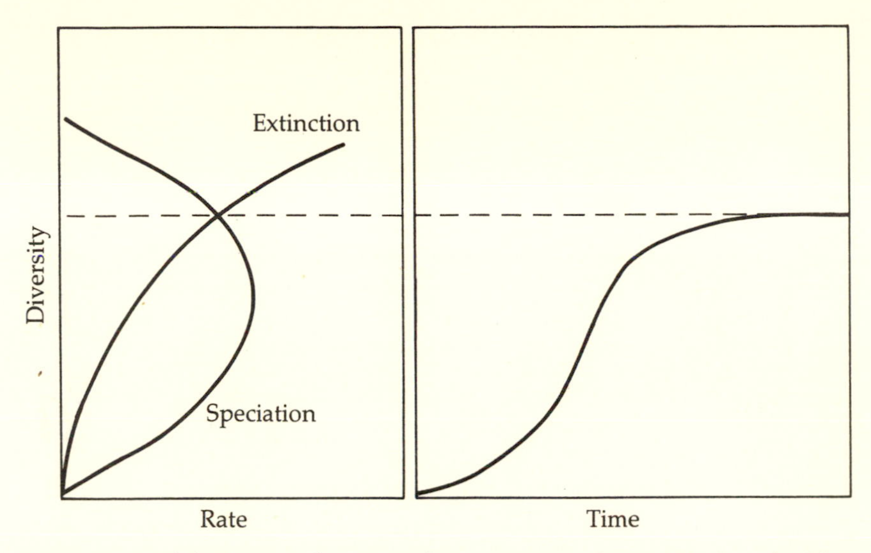

Figure 14–1. If the rates of origin of new species (speciation) and of loss of old species (extinction) depend on diversity as indicated on the left, diversity (the number of species) will increase from an initially low value to an equilibrium as shown on the right.

Consider the expansion and diversification of the metazoa in the late Precambrian and early Paleozoic. Because these were the first metazoa, diversity was initially low and ecological opportunities were profuse. An exponential increase in diversity resulted as new species arose from older species and in turn gave birth to still newer species to occupy a world of vacant ecological niches. Increasing numbers of species gradually filled these niches, however, so that the probability of success of a new species diminished and the per species rate of speciation declined. At the same time the per species rate of extinction increased with the increasing competition that resulted from increasing numbers of species until speciation and extinction rates became equal and diversity reached an equilibrium value.

A careful analysis of the diversity of marine metazoan orders by J. John Sepkoski of the University of Chicago reveals just this behavior in the late Precambrian and Cambrian (650 to 500 million years ago). Interestingly, the data reveal a second period of growth in diversity beginning in the Ordovician and leveling off at a higher equilibrium value in the Silurian (see Figure 13–1).

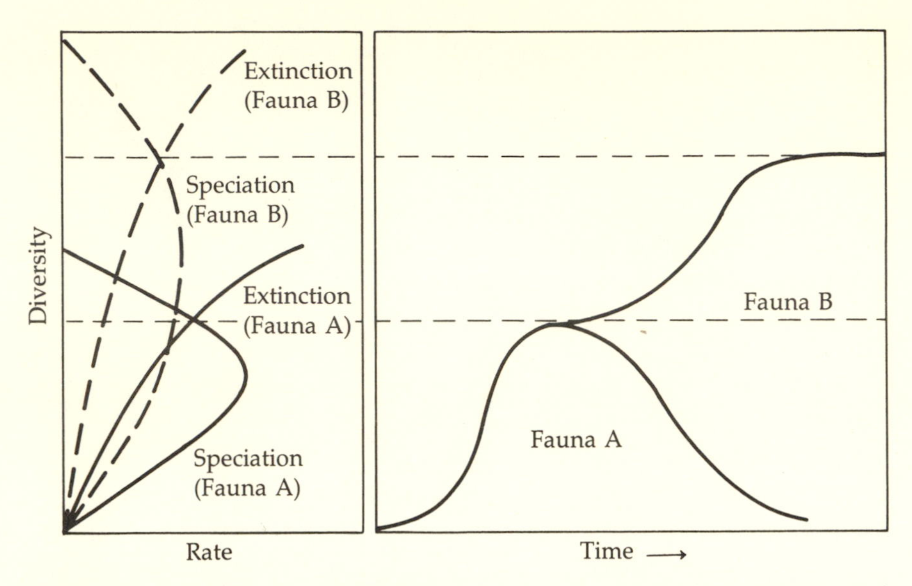

Figure 14–2. Fauna A (Cambrian) with larger rates of speciation and extinction flourishes at first, but is gradually replaced by Fauna B (Paleozoic), which expands more slowly but reaches equilibrium at a larger total diversity.

Sepkoski has shown how these two periods of growth and their different equilibria can be understood in terms of two faunal groups with different speciation and extinction rates, both dependent on the total diversity of the two faunas combined (Figure 14–2). The first, a "Cambrian" group, was characterized by relatively large per species rates of speciation and extinction and relatively strong dependence of these rates on diversity. It achieved an initial equilibrium in diversity before the end of the Cambrian. Meanwhile, the second, "Paleozoic," fauna was diversifying more slowly, its expansion being retarded by competition with the more rapidly radiating Cambrian fauna. By Ordovician time, however, the Paleozoic fauna had become sufficiently diverse to affect speciation and extinction rates of the Cambrian fauna, which began a slow decline in diversity. This decline was more than offset by continuing expansion of the Paleozoic fauna so that total diversity climbed to a new and higher equilibrium value by the end of the Ordovician. This second equilibrium contained more than twice as many families as the earlier Cambrian equilibrium, and most of these families were members of the Paleozoic fauna (Figure 13–1).

The Cambrian fauna may reasonably be described as generalists, not too closely adapted to the environment and able to exploit a wide range of resources. Because of their unspecialized requirements, generalists can be expected to expand rapidly into an essentially vacant ecosystem. Nevertheless their broad ecological niches imply that a relatively small number of generalist species can coexist, so equilibrium is achieved at a relatively low diversity. The more specialized Paleozoic fauna took longer to develop, but achieved a higher diversity by partitioning ecospace into narrower ecological niches. Specialists can make more efficient use of limited resources than generalists and so can be expected to drive the generalists to extinction, at least in stable environments, as the carrying capacity of the biosphere is approached.

Sepkoski's analysis revealed a third great wave of diversification that began at the end of the Paleozoic (see Figure 13–1). At that time the Paleozoic fauna was gradually replaced by a modern fauna, and the total number of marine metazoan families rose to a new high, approximately twice as large as the Paleozoic equilibrium. It is not clear whether this expansion was a consequence of further specialization. That it was not simply a result of increasing provinciality is shown by the variation with time in the numbers of species occupying particular habitats.

Data on a large number of fossil communities of bottom-dwelling marine animals have been analyzed by Richard K. Bambach of Virginia Polytechnic Institute and State University. He classified the communities in terms of the variability of their environments and then determined how diversity in each environment has changed over time (Figure 14–3). The data show no significant change in the diversity of organisms living in highly variable environments such as tidal flats and estuaries. In the most stable environments, however, diversity increased by 50% after the Cambrian and Ordovician, remained approximately constant through the rest of the Paleozoic and Mesozoic, and then doubled in the Cenozoic. Diversity in the stable environments—open sea below the level of disruption by storms—has always

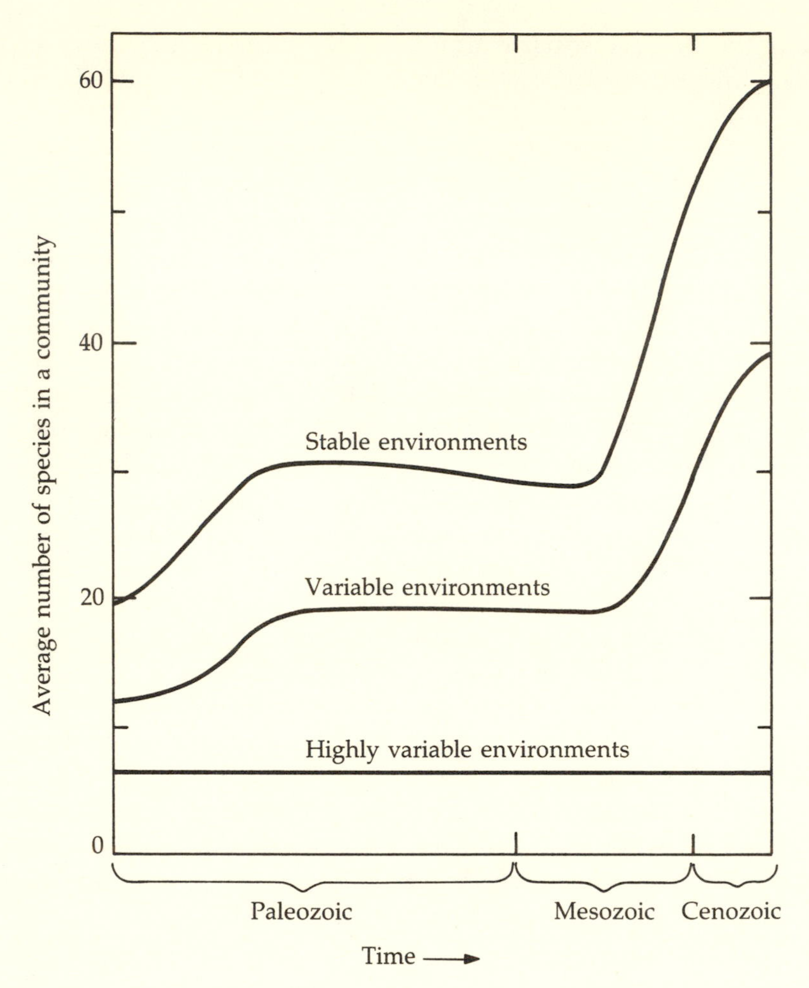

Figure 14–3. Changes over time in the average number of species per community of bottom-dwelling marine animals.

been much higher than in variable environments, an observation consistent with our earlier discussion of the influence of environmental stability.

Thus the total diversity of life can change both as a result of changes in the diversities of communities within particular habitats and as a result of changes in the number of distinctly different communities (provinciality). The increase through time in the diversity within habitats must be largely a consequence of biological evolution and changes in the ecological structure of communities. In contrast, changes in provinciality result mainly from geological processes, particularly continental drift with its impact on geography and climate.

The fossil record yields data not only on diversity through time but also on the rates of extinction and of speciation. Broad features of the histories of these rates have been known for some time, but the pattern has been brought more sharply into focus in a recent analysis by Sepkoski and David M. Raup of the Field Museum of Natural History (Chicago) of the rate of extinction of marine animal families. From statistical study of many data they find that there has been a fairly steady background rate of extinction that has declined slowly from a rate of about 5 families per million years in the Cambrian to a recent rate of about 2 families per million years. They speculate that this decline reflects increasing fitness of the biota as a consequence of continued biological evolution. Superimposed on the uniform background extinctions are five extinction events when, for geologically short periods of time, the rate of extinction was markedly above the background (Figure 14–4). Of these, the mass extinctions that marked the end of the Paleozoic and Mesozoic eras have attracted the most interest.

In the so-called Permo-Triassic event, 225 million years ago, half of all families of marine animals became extinct. This is the time when the continental aggregation of Pangaea was close to its fullest development. I have already discussed how the formation of Pangaea should have led to reduced provinciality and decreased environmental stability. Both factors could have contributed to the extinctions by causing competition between species that had previously been isolated by geographical or ecological factors. Another factor that appears to have been important is sea level. Sea level relative to the continents affects the total area of shallow water open to colonization by marine animals. It has fluctuated during the course of geologic history as a result of tectonic processes described below. The end of the Permian was a time of particularly low sea level and particularly restricted areas of shallow sea. The correlation between area and biological diversity has already been mentioned. Decreasing area at the end of the Paleozoic may have contributed to the Permo-Triassic extinctions by crowding species together, causing more competition and predation.

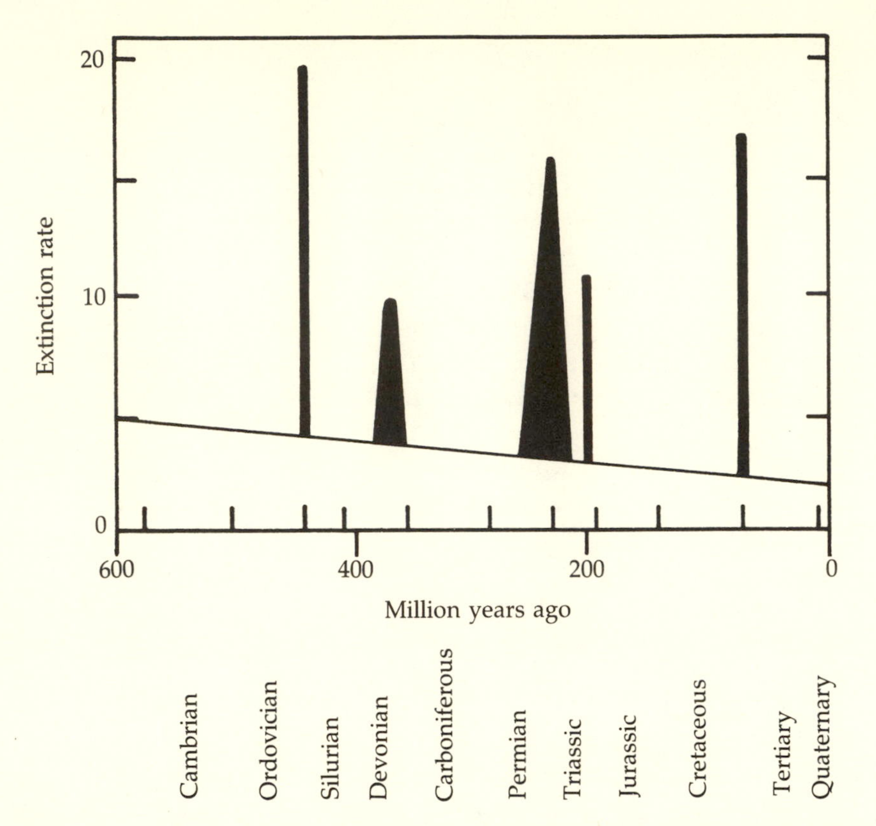

Figure 14–4. The rate of extinction of marine animal families per million years shows five sharp peaks superimposed on a gradually declining background.

Sea-floor spreading influences sea level because the mid-ocean ridges that form where new sea floor is being generated occupy a volume of the ocean basins that would otherwise be filled with water (Figure 14–5). The ridges are elevated because they are composed of hot rock that is less dense than the surrounding oceanic crust. As new crust moves away from the crest of the ridge it cools and contracts, and its elevation decreases—the depth of the sea floor increases. This transition covers a larger area where spreading is rapid, so the ridges produced by rapid spreading displace a larger volume of sea water than ridges with low spreading rates. Episodes of rapid spreading or a spreading axis of great length therefore correspond to high sea level. If sea floor spreading were to stop altogether, the ridges would disap-

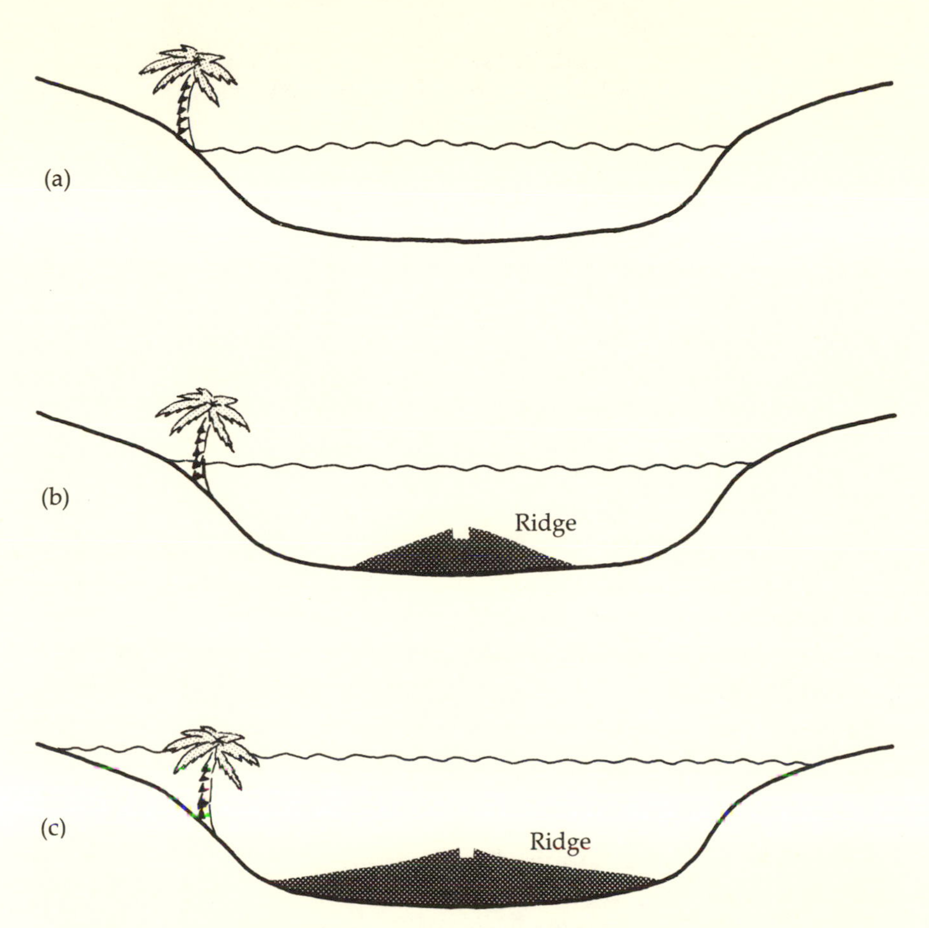

Figure 14–5. The volume of a midocean ridge displaces sea water onto the land (compare parts a and b). The volume of a rapidly spreading ridge is larger than that of a slowly spreading ridge (compare parts b and c), so times of rapid spreading correspond to times of high sea level and extensive shallow seas.

pear as they cooled, and shallow seas would be drained by water flowing into the ocean basins to occupy the space previously occupied by ridge. Low sea level at the end of the Paleozoic was probably a result of a temporary reduction in sea floor spreading activity, which may in turn have resulted from the continental collisions that led to the aggregation of Pangaea.

There is a measure of agreement that the Permo-Triassic extinctions were caused by geographic change, though the relative roles of provinciality, stability, and area remain subject to debate. Enhanced speciation followed the extinctions as the environmental constraints relaxed during the Triassic and the biota expanded in response to ecological opportunities left unexploited by the extinctions. The record shows a clear lag between enhanced extinction rate and enhanced speciation rate, so there is little likelihood that the extinctions were the result of displacement of old species by new species. An example of such biologically caused extinction is the gradual decline of the Cambrian fauna in response to increasing diversity of the Paleozoic fauna.

Still another mode of extinction seems to be illustrated by the Cretaceous-Tertiary event (65 million years ago). This is by no means as large as the Permo-Triassic mass extinction. Only 11% of marine animal families became extinct at the end of the Mesozoic. Nevertheless, this extinction may have had a particularly significant impact on the evolution of life because it was at this time that the dinosaurs, already in decline, finally died out and the way was opened for the adaptive expansion of mammals. Geochemical evidence has very recently been interpreted to indicate that a large extraterrestrial body, either a meteorite or a comet, struck the earth at the end of the Cretaceous. Its debris has been found in sediments deposited at this time in widely scattered areas of the globe. It has been suggested that this impact killed many of the phytoplankton that are the primary producers of organic matter in the ocean, though the mechanism of this mortality is not yet known. Productivity recovered quite slowly in the surface layers of the ocean, and even those Cretaceous species that survived the mass mortality became extinct in the early Tertiary and were replaced by more modern lines.

Death of the phytoplankton disrupted the food supply for marine animals, leading to extinctions in this domain also, and the drop in the rate of photosynthesis may have upset the chemical balance of the ocean, leading to deep waters more corrosive to calcium carbonate and a rise in the partial pressure of carbon dioxide in the atmosphere. Climatic changes resulting from the changing chemical balance of ocean and atmosphere may have contributed to extinctions on land, particularly the extinction of many reptilian lines, including the dinosaurs. Why the mammals suffered so little is one of the many questions concerning this extinction event that still require elucidation. Research in this area is vigorous, and continued rapid progress can be expected. Apart from its historical importance, the Cretaceous-Tertiary extinctions may provide indications of how the biosphere could respond to perturbations introduced by man. An increase in atmospheric carbon dioxide caused by the burning of fossil fuels is one such perturbation, and poisoning of the ocean or a very large explosion are others.

Suggested Reading

Smith, J. M., Ed. *Evolution Now.* San Francisco: W. H. Freeman, 1982.

Stanley, S. M. *Fossils, Genes, and the Origin of Species.* New York: Basic Books, 1981.

CHAPTER 15

The Age of Mankind

THE STORY of mankind provides perhaps the most pertinent example there is of the interdependence of life and the environment. The story divides naturally into three phases. The first was a phase of biological (genetic) evolution in response to environmental pressures. Mankind's story, up to this point, is no different from that of any of the myriads of other species that have inhabited the planet. The second phase, however, is unique to humans. It involved the evolution of culture as a means of dealing with the environment. The third phase is comparatively recent, but it marks the time when the story of mankind and the story of the biosphere became one and the same. Mankind is now an agent of environmental change on a global scale almost as powerful as the prokaryotic microbes that flourished in Precambrian seas. This chapter describes the origin and evolution of humans and their culture in the light of the principles developed in earlier chapters and applies these principles to a consideration of human impact on the biosphere.

Mankind's earliest primate ancestors, the prosimians, appeared about 70 million years ago (Figure 15–1). Like other early mammals they were originally insectivores, but they probably became omnivorous during the course of an arboreal way of life. The lemurs of Madagascar are modern representatives of the prosimian group. Adaptation to a life in trees led to two changes in the skeletal organization of the prosimians that were of great importance in the subsequent evolution of primates. The first of

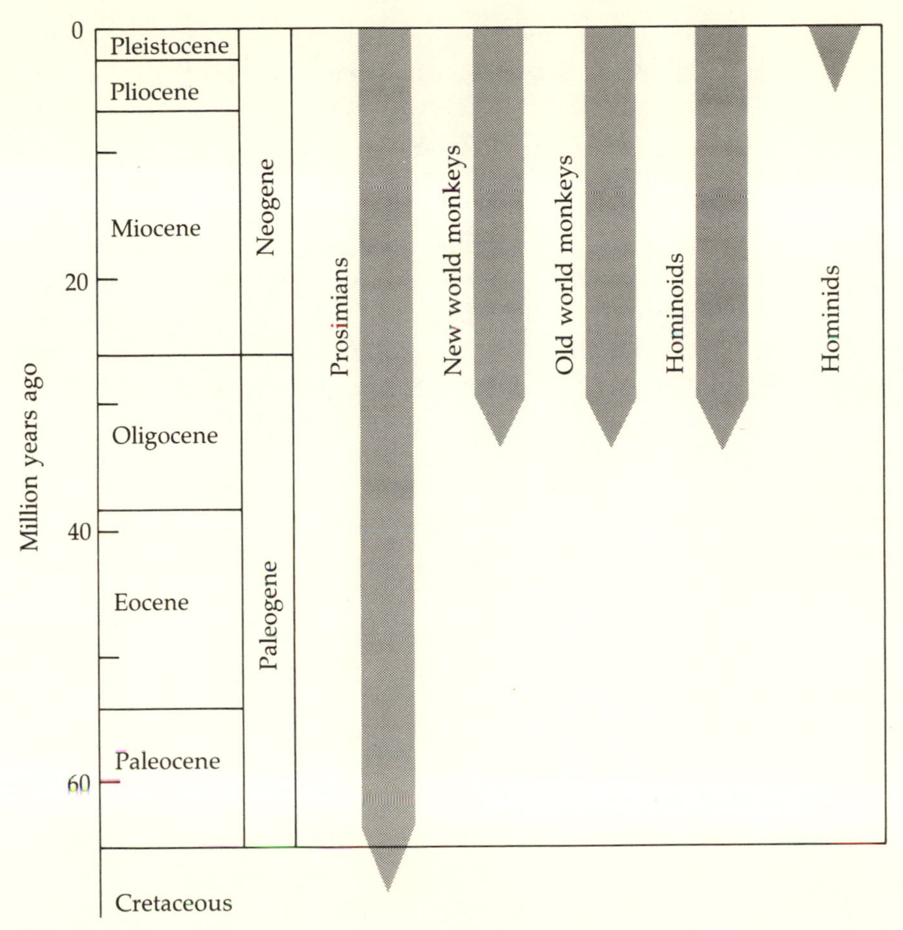

Figure 15–1. Evolutionary developments leading to the origin of humans.

these was the grasping hand with opposable thumb, which was originally an adaptation for holding onto limbs and branches but was later to permit carrying and the use of tools. The second was the development of stereoscopic vision, which was also important to subsequent manipulation of objects with the hands. In most mammals the eyes are placed toward the side of the head. This arrangement allows a maximum field of view but provides for little overlap between the fields of each eye. Primates, however, needed to judge the distance to the next branch precisely. In the course of time their eyes moved forward and their faces flattened.

About 35 million years ago three groups of large, more advanced primates arose from prosimian stock. These became the new-world monkeys of South America, the old-world monkeys of Africa and Asia, and the hominoids, which evolved into chimpanzees, gorillas, orangutans, gibbons, and humans.

The evolution of the hominoids reflects increasing adaptation to life on the ground. The first hominoids, called dryopithecines, stood two or three feet tall and lacked a tail. In the early Miocene, about 20 million years ago, they lived mainly in tropical forests, eating soft fruits and insects. By the middle Miocene, about 15 million years ago, dryopithecine apes were flourishing in country with open woodland and savannah vegetation. Their teeth and jaws reveal adaptations to a diet including small, tough foods like nuts and roots. This change may have been stimulated by the global decline in temperature that culminated in the ice ages. As tropical forests gave way to temperate zone vegetation with a more marked seasonal variation, the large apes may have found that the food available to them in the trees was inadequate. They became more adept at life on the ground as they spent more time there looking for food and getting from one tree to another.

The first representative of the hominid line, *Australopithecus,* appeared about 4 million years ago. To date, its remains have been found only in Africa, an area that was relatively free of the environmental disruption that marked the onset of the ice ages, but it may well have been more widely distributed.

Reconstruction of the paleoenvironments of the various sites at which *Australopithecus* has been discovered provides some indication of its way of life. The sites were characterized by semi-arid climates marked by alternating wet and dry seasons and the presence of open woodland or savannah vegetation. The sites typically occurred at the interface between two different types of environment—between open and closed vegetation, for example, or along a lake shore or stream. These settings provided the australopithecines with diverse surroundings that encouraged adaptability and experimentation with new sources of food and new ways of life.

A particularly significant adaptation was an upright stance and bipedal locomotion. This adaptation was essentially complete by about 4 million years ago, a conclusion that has been spectacularly confirmed by the discovery of preserved hominid footprints at Laetoli in Tanzania. The expansion of the hominid brain apparently occurred considerably later, around 2 or 3 million years ago, when the oldest recognizable tools also appear in the fossil record.

Walking on two legs left the hands free, making it easier to carry and use tools, but the time lag just referred to suggests that the two capabilities did not evolve together. What selective pressures may have led to the evolution of bipedal locomotion? C. Owen Lovejoy of Kent State University has argued for the importance of pair bonding and child rearing in this connection. Advanced primates have a longer life span than primitive ones, and their physical development takes longer. A long period of dependent childhood provides ample opportunity for teaching children how to survive. This promotes the evolution of culture, but it also places a burden on the parents who must feed and protect the child. Single births fairly widely separated in time mean that reproductive success requires a low infant mortality.

Lovejoy suggests that hominids dealt with these problems by leaving mother and child in a secure location while father went out to forage for food. Birds and other animals use a similar strategy of parenting, for much shorter periods of time, but the hominid diet and physiology does not permit enough food to be carried in the mouth or regurgitated. Instead, hominids were under selection pressure to carry food in their hands. The first tool may have been a sort of basket. As opposed to standing up occasionally to look around, hand carrying required extended periods of upright posture. The anatomical changes that made this stance less tiring would therefore have been favored.

The subsequent evolution of humans included increases in the size of the brain probably associated with the use of tools. The emphasis shifted from biological evolution to cultural evolution. The evolution of culture is intimately related to the ice age fluctuations in climate that have characterized the last 2 million years of earth history (Figure 15–2). It is therefore appropriate at this point to consider climate change and some of the mechanisms that cause it.

There is abundant evidence of extensive glaciation in North America and Europe during the recent geological past. Long Island, for example, is composed of the debris deposited at the edge of one of the continental ice sheets, and the U-shaped basins

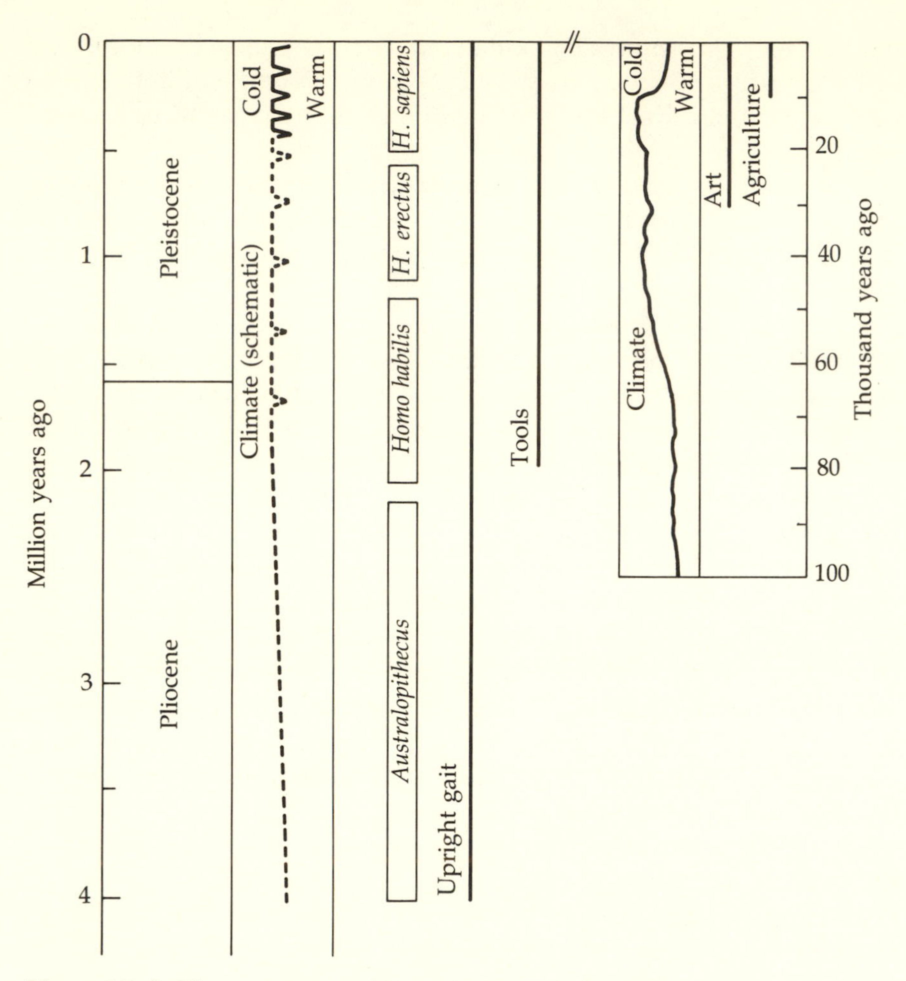

Figure 15–2. The ancestors of *Homo sapiens,* their culture, and some indications of changing climate.

of the Finger Lakes in upper New York were carved by the southward movement of the ice. The recent past has also been a time of profound climatic instability, in which cold glacial periods have alternated with warm interglacial periods. At least four major periods of glaciation have been recognized. The most recent one began about 65,000 years ago and ended only 10,000 years ago. At its peak, ice covered perhaps half of North America and a substantial part of Europe. Superimposed on these major oscillations have been fluctuations of smaller amplitude and shorter period during which glaciers alternately advanced and retreated.

Climate during most of geological history has been markedly different. There were similar periods of extensive continental glaciation about 250, 450, 700, and 2300 million years ago, but glaciers were generally restricted to high mountains and possibly to polar regions during the long periods of time that separated these episodes. The mechanisms responsible for the recurrence of ice ages and for the fluctuations in climate that occur within them are understood only in general terms.

One possible cause of change in average global temperature is change in the brightness of the sun. Careful historical research by John Eddy of the High Altitude Observatory in Boulder, Colorado, has recently established that at least some properties of the sun have varied with time. Drawing on discoveries first made by F. W. G. Spörer and E. W. Maunder in the 1890s, Eddy has found that the frequency of occurrence of sunspots (associated with disturbances on the surface of the sun) has not been constant during the time since regular telescopic observations began. During a 70-year period between 1645 and 1715, for example, there were practically no sunspots at all. This Maunder minimum in sunspot activity and the preceding Spörer minimum around 1500 correspond closely to a time of harsh winters accompanied by the advance of Alpine glaciers that is known as the Little Ice Age. Physical mechanisms have been proposed that relate the frequency of occurrence of sunspots to the total output of energy by the sun, so it is quite likely that little ice ages, with durations of about a hundred years, result from solar variability. The record of solar variability does not extend back far enough, however, to reveal fluctuations of longer period. Longer period fluctuations of climate are generally attributed to other causes.

Change in the brightness of the sun is not the only possible cause of change in average global temperature. A change in the reflectivity of the earth (called the albedo), which determines what fraction of the incident solar energy is absorbed, is another possibility. Snow, of course, has a high reflectivity, so there is a tendency for an ice age to reinforce itself. As glaciers spread, they reduce the amount of solar heat absorbed. Clouds also have a high albedo. This kind of feedback contributes to the difficulty of gaining a quantitative understanding of climatic change.

In an effort to provide a firmer observational foundation for studies of the problem, a group of scientists concerned with climate change have organized the CLIMAP project (Climate: Long-range Investigation, Mapping, and Prediction). As part of CLIMAP, James Hays of Columbia University, John Imbrie of Brown University, and N. J. Shackleton of Cambridge University conducted a detailed study of indicators of past climate preserved in cores recovered from deep sea sediments. The study revealed remarkably regular oscillations in climate during the last 350,000 years, with periods of 23,000, 42,000, and 100,000 years.

These regular oscillations provide strong support for a theory of climatic change first suggested in 1830 by the astronomer John Herschel. This theory is now called the Milankovitch theory in honor of Milutin Milankovitch, a Serbian scientist who published a detailed exposition of it in 1941. It depends on regular variations in the earth's orbit about the sun and in the orientation of the earth's axis of rotation that result from gravitational interactions within the Solar System (Figure 15–3). The ellipticity of the orbit, which determines the difference between the greatest and least distances of the earth from the sun during the course of a year varies in a cycle of 105,000 years. The tilt of the axis of rotation with respect to the plane of the orbit, which governs the summer to winter contrast between solar energy input at a given latitude, varies in a 41,000-year cycle. Finally, the orientation in space of the rotation axis traces out a circle about the normal to the orbital plane every 21,000 years. This orientation determines whether it is northern or southern summer when the earth is closest to the sun. The close agreement between the periodicities predicted by the Milankovitch theory and the periods detected in the climatic record of the past few hundred thousand years provides strong evidence of the importance of orbital change in causing recurrent advances and retreats of the glaciers during ice ages.

The Milankovitch theory does not, however, explain why ice ages have occurred so infrequently during earth history. One of the most plausible explanations of this phenomenon derives from the work of Alfred Wegener more than half a century ago. Wegener argued that it is easier to imagine continents moving than to imagine polar climatic zones at low latitudes (near the equator).

A climatic potential for continental glaciation exists and presumably has always existed at high latitudes. This potential has

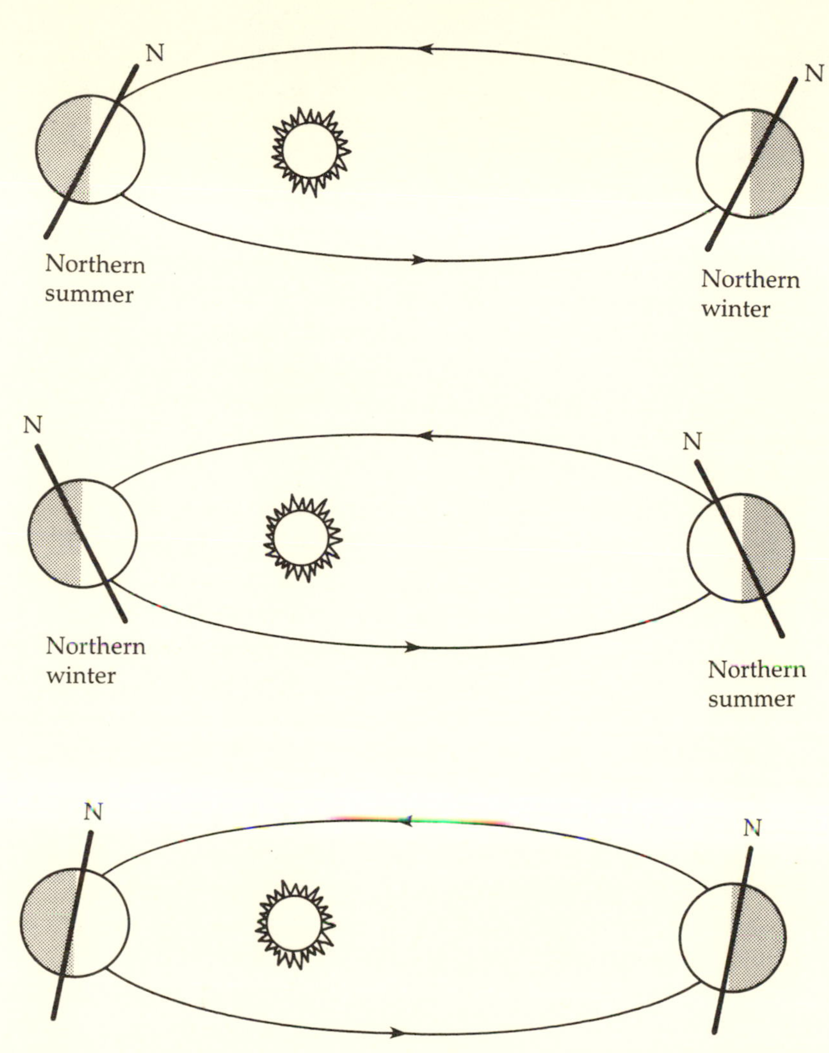

Figure 15–3. Changes in the earth's orbit and their effect on climate. Because the earth's orbit is elliptical (exaggerated in the drawing), the distance of the earth from the sun varies during the year. The effect (top) is to make the northern hemisphere winter colder than the southern hemisphere winter. This situation is reversed after 10,500 years (center) because the earth's axis of rotation traces out a circle about the normal to the orbital plane. But the angle between the axis of rotation and the normal to the orbital plane also changes. Seasonal contrasts are much reduced after 20,500 years when this angle is small (bottom). The ellipticity of the orbit varies with a half-period of 52,500 years.

been realized in the form of widespread continental glaciation only when continental drift has carried extensive areas of land into the glacial zone. There is a concentration of land at high northern latitudes today, and there was a similar concentration between 450 and 250 million years ago when the supercontinent of Gondwana was drifting across the South Pole (see Figure 13–4). The southern continents (South America, Africa, Antarctica, and Australia) bear traces of the Paleozoic glacial ages.

Once again there emerges a major role for plate tectonics and continental drift in guiding the course of earth history. Changes in Earth's orbital parameters and the brightness of the sun may be responsible for climatic fluctuations of relatively short period, but according to this theory of climate change, the major disruptions of the biosphere associated with extensive continental glaciation occur only when continents have drifted into the high latitudes in which snow and ice can accumulate.

If (as I shall suggest in the next chapter) the evolution of culture and intelligence was stimulated by the climatic instability of the ice age, it is not inappropriate to wonder what would have happened if continental drift had not resulted in conditions favorable to glaciation relatively early in the age of mammals (mammals replaced reptiles as the dominant land animals only 65 million years ago) or if the configuration of the Solar System had not been such as to cause the recent fluctuations in climate on a time scale corresponding to thousands of human generations. Would intelligent life have appeared on Earth without this apparently fortuitous conjunction of circumstances? This question cannot now be answered.

Suggested Reading

Geophysics Study Committee. *Climate in Earth History.* Washington, DC: National Academy Press, 1982.

Skinner, B. J., Ed. *Climates Past and Present.* Los Altos, CA: William Kaufmann, Inc., 1982.

Tattersall, I. *Man's Ancestors.* London: John Murray, 1970.

CHAPTER 16 *The Age of Ideas*

THE EVOLUTION of *Homo habilis, Homo erectus,* and ultimately *Homo sapiens* from Australopithecine stock occurred before and during the course of the ice ages (see Figure 15–2). The fossil record of these ancestors is meager, but they left a cultural record in the form of the Acheulian (about 1.5 million to about 100,000 years ago) assemblage of stone tools. These are large tools, mostly handaxes and cleavers suitable for cutting, chopping, piercing, scraping, and pounding. They were probably used for butchering game and for the preparation of simple animal and vegetable products. Archeological evidence indicates a predominance of large game in the diet, but vegetable foods were also used. Compared with their predecessors, the Acheulian hunters (who were usually representatives of *Homo erectus*) apparently utilized a narrower range of foods and favored the most easily bagged forms of big game in any one area.

The most remarkable aspect of Acheulian culture is its stability both in time and in space. It persisted with very little change from 1.5 million to 200,000 years ago in East Africa and from 500,000 to possibly 100,000 years ago in western Europe. The differences between Acheulian tool kits from Africa and from Europe are smaller than the differences between 10,000-year-old tool kits from adjacent European valleys. Thus, by modern standards the Acheulians were extremely conservative, preserving their way of life with little change for tens of thousands of generations. Presumably, their manipulative, cognitive, and organizational skills were low, because of biological limitations, so their rate of cultural evolution was correspondingly slow.

Their reliance on a diet of big game meant that fewer Acheulians could be supported in a given area than had been the case for the Australopithecines. Mobility was important as they followed their prey. They dispersed widely, but showed a preference for open, grassy environments with large herds of gregarious herbivores. A rich and dependable supply of game was more important to them than temperature, for example. They were equally at home in the tropical savannahs of East Africa and the mountain grasslands of glacial-age Spain. The glacial periods, in fact, were favorable for the Acheulian hunters because they opened up substantial areas of tundra and light woodland in Europe, permitting large herds of herbivores to develop. The forests were thicker and more continuous during the warm interglacial periods, so there were fewer large herbivores and therefore fewer hunters.

There is evidence that the Acheulians lived in small groups consisting of no more than a few tens of individuals. In times of favorable climate, when game was abundant, the population would multiply, and groups would spread out over wide areas. The population was sufficiently sparse that encounters between groups may have been exceedingly rare during these periods of dispersal. The isolation and the small size of the group led to inbreeding and fairly rapid change in the genetic make-up of different groups. The fossil record, indeed, reveals considerable variability in the physical types represented among the Acheulians. When unfavorable climatic developments caused a decline in big game, the hunters were forced to abandon marginal areas. Survivors gathered in places where the supply of food and water was most reliable. Contact between groups at these times permitted interbreeding, causing advantageous adaptations to spread through a larger population. This alternation of periods of isolation of small breeding groups with periods of exchange of genes between groups could have hastened the processes of biological evolution leading up to the origin of *Homo sapiens.* The cyclic variation in the availability of resources promoted natural selection by alternating periods of expanding population (population flushes) with periods of ruthless competition.

The climatic fluctuations of the ice age may therefore have

accelerated biological change as well as favoring the evolution of ingenuity and adaptability. By about 35,000 years ago, human culture had evolved to the point of cave art. At about the same time humans were able to voyage over considerable stretches of open water to colonize Australia from Indonesia. In a fairly short span of time early humans also managed to colonize the Americas, spreading all the way to Tierra del Fuego and, in the process, adapting successfully to almost all the environments the world has to offer.

The end of the last glacial period also brought about the decline of the hunting life. Warmer climates in Eurasia and North America and wetter climates in the tropics caused the spread of forests at the expense of open grassland. Gathering of plant products as a source of food increased in importance relative to hunting. In time, humans began to make their tasks a little easier by gathering desirable plants in one area and keeping undesirable plants away. This practice, the beginning of agriculture, is thought to have started in tropical rain forests and probably involved the root crops that still form staples in these areas today. Because root crops are generally deficient in protein, it was necessary to supplement the diet with animal protein obtained by hunting and fishing.

A more balanced diet was provided by seed crops such as rice and grain. As humans learned to cultivate and harvest these crops their dependence on animal protein lessened, permitting agriculture to spread to areas where opportunities for hunting and fishing were limited. Domestication of animals increased mankind's independence of natural ecosystems. Settled communities became possible, and a relatively assured supply of food allowed populations to grow. As agriculture became more productive, urban centers developed in which specialized craftsmen were able to devote themselves to activities such as pottery, carpentry, and metal working. The increasing concentration of the population along with a diet of agricultural products led to further biological evolution, favoring smaller teeth as well as resistance to the diseases associated with high-density living. What came next is recorded in history books, another important product of cultural evolution.

Culture mediates between society and the environment, and cultural evolution has been interwoven with biological evolution throughout human history. Environmental changes during the ice ages triggered cultural responses that favored biological change that in turn permitted further evolution of culture. Culture has permitted humans to adapt to a wide range of different environments. At the same time, the bond of common culture has maintained the biological unity of the species in spite of a diversity of habitats and ways of life.

In fact, cultural evolution has been so successful that humans, alone among metazoa, are now able to change the entire biosphere. The factor above all others responsible for this impact is growth in population. Remember that one of the basic mechanisms of evolution is for organisms to produce more offspring than can be supported by available resources, so that only the individuals most suited to their environment are likely to survive to provide genetic material to the next generation. This mechanism seems like a good thing for clams and eels, but is less attractive when we think of it applying to human beings. As early as 1798, Thomas Malthus pointed out that human population has usually been brought into balance with resources by famine, epidemic, and war. Will this still be true in 1998?

Population growth rates today are largest in the less developed countries. In most of Europe and North America the annual rate of growth of population is less than 1%, while in Africa, Asia, and Latin America the rate ranges from 2.5% to 3%. At an annual growth rate of 3%, population will double in only 24 years. It is doubtful that the food supply will also double.

Three factors contribute to change in the size of a population. They are births, deaths, and migration. Examples of important migrations are very familiar; I will consider the other two factors. Prior to 1700, the rates of population growth in most European countries were as low as they are today, but birth rates and death rates were both high. Death rates decreased sharply as a result of improvements in sanitation, medicine, and nutrition, and population rose rapidly. In time, birth rates fell to levels only a little above death rates. The change from slow growth accompanied by high birth rates and death rates to slow growth with low birth and death rates is called the demographic transition (Figure 16–1). The lower birth rates are presumed to be associated with a higher standard of living.

The less developed countries have achieved the first stage of

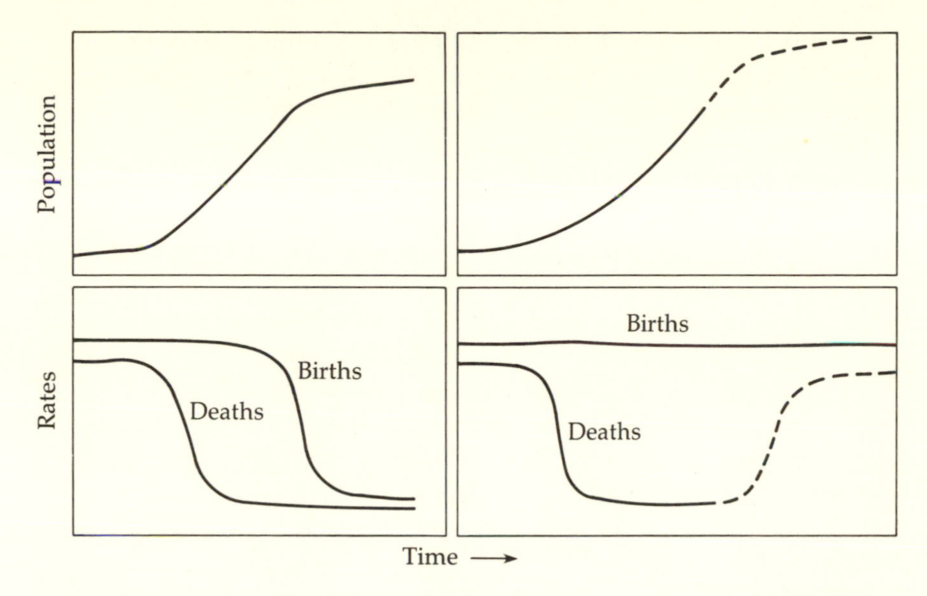

Figure 16–1. Over the last few centuries the industrialized nations have experienced the demographic transition shown on the left. If birth rates do not decline in the less developed countries, population might be stabilized by an increase in death rates as shown on the right.

the demographic transition. Death rates have dropped abruptly but birth rates have not, leading to the present explosive growth in population. It is not clear that the rate of economic growth of these countries can keep pace with the growth of population. The developed countries passed through the demographic transition at a time when fuel and mineral resources were abundant and therefore cheap. Economic growth was further stimulated in many of them by the products and markets of colonial empires. Today the inhabitants of less developed countries, in contrast, are trying to raise their standards of living in a world no longer possessed of seemingly limitless supplies of energy and mineral wealth.

If standards of living do not improve, will the less developed countries ever reach the stage of declining birth rates? If birth rates do not decline, then population will presumably be stabilized by an increase in the death rate. So it is by no means clear that the less developed countries will achieve the demographic transition. Their failure to do so would mean trouble for rich and poor alike.

There is a momentum to population growth that means that population would not immediately stabilize even if parents all over the world began at once to limit their families to the numbers of children needed to replace themselves. The momentum arises because of the distribution in ages of people brought about by population growth. More than one third of the world's population is under 15 years of age. There are therefore far more young people who will soon be adding to population than there are old people about to die. This imbalance between young and old would cause population to climb by 30% or more over a period of 50 to 75 years, even if birth rates were to drop immediately to replacement levels (a little over 2 children, on average, per family). The most optimistic assumptions about when replacement fertility might become the worldwide norm indicate that the population cannot stabilize at a level below 8 billion people, twice the global population of 1976.

Thus there are likely to be a lot more people on Earth in the very near future. What is the impact of people on the biosphere? A familiar answer is that people cause pollution. People introduce new and unfamiliar chemicals into the environment or else release familiar chemicals in amounts much larger than those to which the biota has become adjusted. Some species tolerate the change better than others, so ecosystems are perturbed, possibly even disrupted. This ecological disturbance is perhaps of greater concern than the aesthetic aspects of pollution.

In spite of their technological sophistication, humans still depend heavily on the biological activities of other species. The rest of the biota provides essential services that we cannot duplicate, either because we do not yet understand the basic mechanisms or because we cannot carry out the appropriate activities on a scale sufficiently vast. A prime example of a service to humanity provided by other organisms is the photosynthetic conversion of inorganic nutrients into resources of food and fiber using the energy of sunlight. Other examples include the recycling of waste materials ranging from sewage to industrial effluents (a service reflected in the term "biodegradable"); the pollination by insects of most fruits, vegetables, and berries; and the control of most potential pests by natural enemies. By disrupting natural ecosystems humans may diminish the carrying capacity of the biosphere.

Even without pollution, the efforts of mankind to provide for a burgeoning population would diminish the diversity of the planetary biota. An undisturbed ecosystem, be it woodland, forest, or prairie, supports a much greater variety of plants, insects, and animals than a field of corn, for example, or a suburban shopping center. Modern agriculture depends for its productivity on the maintenance of unnaturally simple ecological communities. The ideal is a large area covered by a small number of crops. Weeds and pests are ruthlessly controlled. As increasing areas of the globe are ploughed or paved over, the area occupied by natural ecosystems is diminished. Ecological research has shown that the number of different species in an ecosystem decreases as its area decreases, so the elimination of natural ecosystems means a decrease in biological diversity. It is estimated that 200 species of plants disappear every year, mainly in the tropics, where highly diverse forest ecosystems are cleared to make way for agriculture.

Further inroads on diversity result from human mobility. In the course of their restless travels around the globe people break down provincial boundaries by transporting organisms, deliberately or inadvertently, from one place to another. Endemic populations frequently suffer as a result of competition or predation by imported species. Examples are the extinction of the dodo and the decimation of North American elm trees by the Dutch elm disease.

Diminution of diversity is of concern in part because it may decrease ecological stability. A stable ecosystem is one without large, rapid change in the sizes of its constituent populations. A complex community can have this kind of stability. If one species of herbivore is eradicated by drought or disease, the carnivores can survive on other kinds of herbivores. If a population of predators dwindles for some reason, there are other predators to control the growth in population of the prey species. A classic example of the instability of a simple ecosystem, by way of contrast, is provided by the Irish famine of the nineteenth century. The people of Ireland were largely dependent on a single crop—the potato. One and a half million deaths resulted when this crop fell victim to a fungus.

An unstable ecosystem, subject to recurrent plagues and famines, is clearly undesirable, but diversity has another important role besides the promotion of stability. A most important function of natural ecosystems is the maintenance of a diversity of genetic information.

Plant breeders develop strains with desirable properties—resistance to disease, tolerance of cold, early yield, and so on—by testing wild strains and experimenting with hybrids. Nature provides a diversity of genetic material with which to experiment, most of it preserved as undomesticated strains in natural ecosystems. So important is this reservoir of genetic material that efforts have been made to conserve it in seed banks. The National Seed Storage Laboratory in Fort Collins, Colorado, for example, has stored the seeds of more than 100,000 different kinds of food and fiber plants. Nevertheless, such stores do not come close to representing all of the genetic variability that presently exists in nature, even for essential crops.

The danger of overreliance on just a few strains of essential crops is that disease can wipe out whole harvests very suddenly. Epidemics of wheat rust and southern corn leaf blight in recent years have demonstrated how vulnerable modern agriculture can be. These epidemics were overcome by the development of new disease-resistant strains by plant breeders working with the genetic variability preserved in wild strains. This breeding potential could be lost as wild strains disappear.

Natural ecosystems constitute a library of genetic information from which can be drawn new and improved crops, new drugs and vaccines, and new biological pest controls. The disappearance of a species or even the decrease of genetic diversity within a species therefore represents the loss of a potential opportunity to improve human welfare. Will the fossil record of the future reveal that the greatest biological crisis of all earth history occurred when *Homo sapiens* overran the globe?

Suggested Reading

Pilbeam, D. *The Ascent of Man: An Introduction to Human Evolution.* New York: Macmillan, 1972.

Scientific American. *The Human Population.* San Francisco: W. H. Freeman and Co., 1974.

Index

*Italic type indicates reference to a figure.
†Boldface type indicates page where term is defined.